DICE WORLD

DICE
WORLD
SCIENCE AND LIFE IN A RANDOM UNIVERSE

BRIAN CLEGG

ICON

First published in the UK in 2013 by
Icon Books Ltd, Omnibus Business Centre,
39–41 North Road, London N7 9DP
email: info@iconbooks.net
www.iconbooks.net

This edition published in the UK in 2014 by Icon Books Ltd

Sold in the UK, Europe and Asia
by Faber & Faber Ltd, Bloomsbury House,
74–77 Great Russell Street,
London WC1B 3DA or their agents

Distributed in the UK, Europe and Asia
by TBS Ltd, TBS Distribution Centre, Colchester Road,
Frating Green, Colchester CO7 7DW

Distributed in South Africa
by Jonathan Ball, Office B4, The District,
41 Sir Lowry Road, Woodstock 7925

Distributed in Australia and New Zealand
by Allen & Unwin Pty Ltd,
PO Box 8500, 83 Alexander Street,
Crows Nest, NSW 2065

Distributed in the USA
by Consortium Book Sales & Distribution,
The Keg House, 34 Thirteenth Avenue NE,
Suite 101, Minneapolis, MN 55413-1007

Distributed in Canada by
Penguin Books Canada,
90 Eglinton Avenue East, Suite 700,
Toronto, Ontario M4P 2YE

ISBN: 978-184831-652-2

Typeset in Melior by Marie Doherty

Printed and bound in the UK by
Clays Ltd, St Ives plc

Contents

CONTENTS

Acknowledgements

For Gillian, Rebecca and Chelsea.

With particular thanks to Keith Rapley and the late John Carney, my mentors when working in operational research – arguably the ultimate discipline for applying probability and statistics to real-world problems.

Many thanks to all those who have helped make this book possible, especially Duncan Heath, my editor at Icon Books, and all the contributions from Andrew Furlow, Dr Peet Morris, Denni Saunders, Dr M.G. Harris, Cathy Murphy, Perry Rees, Katherine Kelly, Paul Tuck, Amanda Lever, Edward Cope, Helen Witney, Dr Henry Gee, Mats Anderson, Liz Warrick, Stacey Croft, Desney Harrington, Kathy Peacock, Sarah Mussi, Matt Brown, Euan Adie, Ethan Friedman, Lynn Price, Amy Cope, Chris Reeves, Diane Rendell, Sue Broughton, Mark Lloyd, David Howkins, Stephen Godden, Dr Harriet Dunbar-Morris, Oriana Morrison-Clarke, Stewart Desson and Henry Lord.

CHAPTER 0

Alea jacta est

The first stage of writing a non-fiction book is usually to sketch out a summary of each chapter that will go in the finished work. At that time, the possibility of writing this chapter had not occurred to me – hence it being chapter zero. The very existence of this chapter is a great example of random influences at work.

Shortly after I began my background research for *Dice World*, I read a book called *The Black Swan* by Nassim Nicholas Taleb. This was not part of my research – it was intended as bedtime reading, light relief from work. I was taking a look at a cult classic that I'd never got around to – I didn't even realise at the time that the book featured chance and probability. It was something of a shock, given that I had already chosen the title *Dice World* for my own book, to find Taleb railing against the image of dice as a cultural reference for randomness. Taleb refers to this as the 'ludic fallacy', his idea being that dice represent the fake, controlled, predictable randomness of games, not the real, wild randomness of life.

To rub this in, Taleb tells his readers a story of two fictional characters faced with a classic game-based challenge that is often used to demonstrate the difficulty many people have with understanding probability. Let's imagine we have a fair coin, which when flipped has a 50:50 chance of coming up heads or tails. On average, we expect the coin to come up heads half of the time and tails half of the time. We flip that coin 99 times in a row and get a head every single time. What's the chance that we will also get a head on the next throw?

In Taleb's story, the first character, an accountant, comes up with the standard 'correct' response you would get from anyone who has a good understanding of the basics of probability. On the hundredth throw, there is still a 50 per cent chance of getting a head. The coin has no memory. To think that somehow you are more likely to get tails this time because heads have come up so many times before is what's known as the gambler's fallacy. There is no connection between the 100th toss and the ones that came before it. You are starting from scratch with a single toss of the coin. The outcome is still 50:50 heads or tails.

The other character in *The Black Swan*, a city trader, thinks that this statistical view on life is rubbish. It's not that he agrees with the gambler's fallacy that the coin is 'due' to come up tails. He has no intention of falling into that trap. In fact, the trader will tell you, there's a high chance instead of getting another head. And he is probably right. Why? Because in the real world, it is much more likely that the person who told us it was a fair coin was lying than it is that you will get 99 heads in a row from tosses of an unbiased coin. The real world is not like a game with nice, easily calculated probabilities and no outside influences. The real world cheats.

That is fair enough as a realistic observation of what the world is like, but I think that Taleb misses the point when it comes to using the image of dice as a device to suggest randomness. The dice are symbolic, they are not a computer model of real-world experience. The way that dice provide an illustration of nature's indifference to our human affairs and desires is an ancient conception that Albert Einstein used as a parable for 20th-century science. Einstein's various statements along the lines of 'God does not play dice' were used to illustrate his frustration with the way that quantum theory suggested the apparently predictable real world was in fact based on unpredictable chance. It was never intended to provide a detailed analysis of the different *kinds* of probability and randomness out there – just a nice picture to illustrate something out of our control. And it works.

The quote that provides the title of this mini-chapter, 'alea jacta est', attributed to Julius Caesar as he crossed the Rubicon and meaning 'the die is cast', illustrates exactly how the throw of the dice provides powerful imagery, in this case identifying a point of no return. To deny dice their figurative role seems unreasonable.

Symbols aren't intended to be reality, and the resonant role of dice and other gaming mechanisms helps us grasp some aspects of probability, even if we have to then extend far beyond this to get a true picture of the real world. Denying their symbolism is a bit like complaining that the road sign showing a landslip only features a handful of rocks where a real landslip may well have millions of bits of debris in it. Symbols are not supposed to be the real thing. So this book proudly remains *Dice World*.

If you are a certain age (which I am), *Dice World* may also bring to mind another book, which was very popular when I was at university – *The Dice Man* by Luke Rhinehart. That novel featured a disaffected psychologist who decides to let dice rule his life, handing every decision he makes over to the throw of the dice to make the choice. For me it was a very sad and depressing novel: it is one thing to be aware of the significance of randomness in our lives – an important thing indeed. But it is something quite different to simply give away all rational choice to a random number generator.

Dice World is not the natural home of the *Dice Man*. Most of us don't make our decisions by throwing dice. But like it or not, randomness can be effectively symbolised by a throw of the dice. And as we will discover, randomness is the underlying heartbeat of our universe.

Welcome to *Dice World*.

CHAPTER 1

Improbable world

The world is a complicated and messy place, especially when you consider the complexities we add to it with our carefully constructed environment. Take a really simple act that most of us perform every day without giving it a thought – switching on an electric light. This is clearly not something we are genetically programmed to deal with from birth. Human beings are pretty well identical to the creatures that evolved to survive on the savannah after their ancestors stopped living in trees over 100,000 years ago. Once you get beyond basic bodily functions and activities, the vast majority of our time in the modern world is spent doing things that the human body did not evolve to do. All the rest of our activities and experiences are relatively newly learned. We live unnatural lives.

It's certainly true that there weren't many light switches 100,000 years ago. So we all have to learn how to turn the light on – and for most of us (until we venture across the Atlantic and find that they incomprehensibly mount their switches the wrong way up on the wall) it is a natural-seeming, easy act. We flick the switch and the light comes on. No real thought involved. It's trivial.

But imagine that you had to program a robot from scratch to switch on the light in your living room. You would need to specify exactly where the switch was located. This would involve providing detail of where each wall was, which wall the switch was on, at what height it was located and at what distance it was from the wall's edge. Alternatively you would need to show your robot exactly what the switch looked like from every possible angle, so the robot could search for it visually. You would also need to specify where and in which direction to apply pressure to the switch, how much pressure to use (it would be embarrassing if the robot snapped the thing off) and when to stop pressing.

What seemed trivial turns out to be anything but a simple task. But more to the point, if you now moved that robot into the hallway and asked it to carry out the same job there, you would have to start all over again. There might be a totally different design of switch with dissimilar physical characteristics. It's highly unlikely this new switch would be in the same place on the wall in the hall as the switch is in the lounge. Set the robot in action without reprogramming it and you would probably end up with a hole punched in the plaster.

As human beings, we simply can't afford the time and effort to do the equivalent of re-programming our brains each time we encounter a different light switch. And so we deal with patterns. We don't learn exactly what each light switch that we encounter is like. Instead we have a broad pattern in mind which specifies 'This is how you switch on a light using a wall switch'. It enables us to recognise the switch in a broad range of styles and then just to do it – press the switch, get the light. Until some clever designer comes up with a switch that works when you speak to it or touch the lamp itself – and then you have to start the discovery process all over again.

Finding patterns

Of course, we didn't evolve an ability to recognise patterns to cope with light switches. But exactly the same flexibility of pattern-matching enables us to spot a predator – or a familiar friendly face – even if we have never been in a particular exact circumstance before, and so to take appropriate action. We work with patterns that give us the ability to reduce the almost infinite set of possible deductions from our sensory inputs to a manageable set we can work with, using the mental shorthand that enables us to just 'flick the switch', 'run from the tiger' or 'see and say "Hi" to Nic.'

We are so good at this pattern-matching that we can achieve it even when we have a surprisingly low amount of information on which to make a judgement – in this we are usually a lot better at filling in the gaps than computers are. This is why the 'CAPTCHA' system, used by websites to ensure that people are taking part rather than software programs, makes use of distorted text with characters that are twisted or run into each other. This is a visual input that a human can usually interpret, but a piece of software struggles with turning into useful data.

Take the three partial pieces of text below:

BANK

DANK

BANK

No one would be challenged to see that the top word reads 'BANK', even though a sizable percentage of the text is missing. We find it trivial to fill in the gaps. In the second example, a whole 50 per cent of the text has been chopped off, but there is still enough there to be sure what the word is. It is only the positioning of the final chop, introducing ambiguity with two possible interpretations of 'BANK' or 'RANK', that finally beats our superb ability to take a partial pattern and reconstruct the whole.

Much of the time this human ability to detect patterns is a real plus. It means that we can work with limited data – and in the real world the set of data that we have available is almost always incomplete. But the danger we face is that the pattern-constructing and -matching systems of our brains are so powerful that we imagine patterns when there is nothing there.

This is a good survival principle. It's better to be sufficiently sensitive that you occasionally see a predator where there isn't one, rather than risk missing a killer that is lurking in the bushes. So we create bogeymen out of shadows and misinterpret all kinds of evidence. We see faces in the shadows, in the clouds, or even in the burn marks on a slice of toast. Pure randomness with no pattern is something we find difficult to relate to – our brains expect to see patterns and they do.

The patterns of science

This pattern-matching isn't just about our low-level, immediate, day-to-day interaction with the environment around us (important though that is). It is also the basis of science. It's strange, in a way, that many of us struggle with science because all the scientific method does is to take the basic mechanism we all use to understand the world without even thinking about it, and formalise that mechanism into a process.

In science we are looking for patterns and rules to explain what the universe and its components do and how they do it. It's as simple as that. The mechanisms modern scientists use may get heavy-duty and scarily mathematical, but the basic principle is still one of looking for patterns. What scientists do is arguably just a simple and rather beautiful formalisation of our natural approach to exploring the unknown.

We start off in a state of ignorance. We gather enough data to be able to formulate a hypothesis about what's going on. Then we test that hypothesis − a kind of predictive pattern − against subsequent observations; if it continues to work, we can build on it. If it fails us, we have to start all over again. That's the scientific method. It should be how we naturally interact with the world too, but all too often, once we get a hypothesis, we get fond of it. We can't let it go despite plenty of evidence to the contrary. And that's when science slips into superstition.

To have any hope of making a scientific approach work, we have to expect some degree of consistency of behaviour from the universe. Take something we think of as a constant, a fixed point of certainty − the speed of light. If this varied from day to day or second to second with no logical reason for that variation, and no way of ever anticipating what the speed will be today, then we could never make use of the speed of light, as astronomers do all the time, to help us understand the universe. Given how much of our exploration of the universe is dependent on light and its speed, this would be totally devastating for cosmology. In fact, without a degree of consistency, the whole concept of science would collapse. We would live in a universe that might as well be magical. It is impossible to draw any hypotheses if every time you do an experiment you get totally different results.

This doesn't mean that there won't be circumstances when the speed of light does vary. We know that it is different in a vacuum from when it is passing through a substance – it is slower in air, still slower in water and so on. There are even substances called Bose Einstein condensates that can effectively bring light to a standstill. This is because photons of light don't pass through matter unaffected but interact with electrons, being absorbed and re-emitted, slowing down their progress. But this isn't a problem for science, because these variations are predictable. I know that the speed of light is different when it's going through space than from when it's going through glass. But for the same medium under the same conditions, I expect to get the same result.

I chose the speed of light intentionally because there is even a theory (a perfectly reasonable theory, though not one with a lot of support at the moment) that the speed of light has not stayed the same over time. According to this theory, over the billions of years of existence of the universe, the speed of light has changed very slightly. If this is true, while it would modify some of our conclusions about exactly what was happening long ago in galaxies far, far away (as they say), it too wouldn't be a huge problem for science, because it is something we could predict and consider the influence of over time.

The randomness confusion

There is, however, one aspect of dealing with reality where our superb pattern-forming skills totally let us down. This is where there genuinely is no pattern; where there is no logic that lets us work out what will happen next; where randomness rules and chaos ensues. The good news is that for many of our basic interactions with the world, repeatability is the name of the game and randomness is under control. But whenever we are dealing with the odds in a game of chance, or the discovery that every aspect of the universe at a fundamental level relies on randomness, we have a serious problem of understanding because our pattern-forming brains start floundering.

Just listen to the victims after a disaster has occurred. They will almost inevitably ask 'Why?' – Why us? Why here? Why now? We all want to find a pattern. We want a reason. But usually with this kind of event there isn't one. The event itself will have a cause, but there is no reason for the 'Why us?' type questions. Just imagine, as sometimes happens, that a child has been struck by lightning, or swept away by a flash flood. I have no doubt that his or her family would be asking 'Why us, when there are so many families who don't have to suffer this?' We struggle so much to accept that any event can be the result of true randomness. Many in the past have invoked wrathful deities to explain an outbreak of sickness in their village, blaming it on the bad behaviour of the inhabitants. Such reasoning doesn't make any sense, but it establishes a pattern.

Even now, in the 21st century, this can happen. In 2010 an Iranian cleric announced to the world that women who wear unsuitable clothing or behave promiscuously are to blame for the incidence of earthquakes. 'Many women who do not dress modestly ... lead young men astray, corrupt their chastity and spread adultery in society, which increases earthquakes,' said Hojatoleslam Kazem Sedighi, according to the Iranian media. Sedighi, who was responding to an announcement from Iran's president Ahmadinejad that Tehran was at risk of being hit by an earthquake, told his followers, 'What can we do to avoid being buried under the rubble? There is no other solution but to take refuge in religion and to adapt our lives to Islam's moral codes.' The pattern is coming to the fore again: the earthquake and the suffering it brings has to have a cause, and obviously it is the behaviour of women that is causing it.

We want a pattern, but so often, everything from dramatic real world events to the weird world of atoms and subatomic particles is governed by a randomness that makes our brains hurt. There may be causes, as there certainly are with earthquakes, but the patterns they form may be impossible to detect with any accuracy because the system involved is too complex and chaotic. Or there may be no cause to an event at all, as is the case with the point in time that a specific radioactive atom decays. Either way, if we decide there is a pattern, we are deluding ourselves.

Weighing up risk

Our inability to deal with chance is illustrated beautifully by the way that, time and again, all of us fail when trying to weigh up risk. Take the following example. Let's say that you hear on the news that a taxi driver has been arrested for attacking a young female passenger. (This happened in my town a couple of years ago.) For the next few weeks, if you have a daughter or know someone who seems a potential victim, you will be reluctant to let them take a taxi. It's human nature, you might say. And it is. But in terms of risk, human nature is getting the response profoundly wrong.

Generally speaking, the risk of being attacked by a taxi driver is very low in the UK. Many thousands of journeys are taken every day without a problem. It is only awareness of the news that has made the risk suddenly seem higher. In fact, the risk has actually just gone *down* substantially. Because the only taxi driver in the area known to attack female passengers is now in custody. It is safer to take a cab than it has been for months – and yet human nature, a commonsense reaction to a scary potential pattern, is to feel that it is more dangerous and to offer to drive your daughter everywhere. This is a good example that makes it clear just how much this problem influences all of us. A statistician will understand perfectly well that there is no extra danger and that the risk is very low. But they are still likely to warn their daughters to be careful in such circumstances. It's not rational, but it is human.

For a second example, let's think of the more large-scale risks that human beings face. When we attempt to estimate risk, we tend to give excessive weight to possibilities with a clear and familiar cause rather than an open, generic risk. Children are far more likely, for instance, to be killed by traffic than by predatory human beings, but we can often focus more on the dangers of paedophiles than on those of traffic because these individuals present a threat that features more in the stories we are told by the media than the less dramatic but much more deadly traffic statistics. Unfortunately for our children's safety, paedophiles make better news stories than traffic accidents.

Randomness confuses the hell out of human beings. Given our dependence on patterns as outlined above, and given that randomness is, in effect, the absence of pattern, we are led inexorably to a difficulty of dealing with randomness. That's a problem, because, as we will discover in this book, different aspects of randomness, chance and probability lie behind most of the world we experience. Patterns are, if anything, an oddity in the universe. Randomness is the norm.

From classical to chaos

There are broadly two types of randomness – the classically random and what I will describe as chaotic randomness. 'Chaos' is a term that is bandied around a lot. I am employing this term – or more specifically the phrase 'chaotically random' – in a more particular sense than the English language term that simply means disordered, but in a broader sense than the kind of chaos theory that was encountered in the movie *Jurassic Park* and that lies behind our inability to predict the weather beyond a few days out.

These two categories, classical and chaotic randomness, are rich in paradox. Individual events in classical randomness are impossible to predict, but the overall behaviour of a collection of classically random objects (like a gas of atoms), obeys rules – a so-called distribution – that makes it relatively easy to decide what will happen to the collection in the future. An obvious example of classical randomness is a typical gambling game (assuming that the game isn't rigged). The outcome should be truly random, but the odds of winning are clear and will predict statistically, over time, what is going to happen. The rules of classical randomness can't tell you the result of a specific game, but they can tell you how the different outcomes should be distributed.

So, for example, taking the totally skill-free game of roulette, there are eighteen black and eighteen red slots on the roulette wheel into which the ball can drop, so if we ignore for the time being the green 0 or 00 slots where the house wins, there should be an $^{18}/_{36}$ (or 1 in 2) chance of winning if you bet on either black or red.

I'll just take a moment to look at the different ways of representing probability, as there are surprisingly many different ways of saying the same thing. If something has a 1 in 4 chance of occurring, then on average, one time out of four tries will produce this result. Think of drawing a playing card from a well-shuffled pack with no jokers. There is a 1 in 4 chance of getting any particular suit. We can also represent this as a fraction. So we could say that there was a ¼ or 0.25 probability of getting, say, a heart. The probability varies between 0: no chance at all (the chance, say of drawing a card from a nonexistent fifth

suit) to 1: will definitely happen (the chance of drawing a card with anything printed on it from the standard
52 cards).

We can also use a percentage, which is just the same
as the fraction, but multiplied by 100. So our 0.25 chance
becomes a 25 per cent probability. In 25 per cent of cases,
for instance, we expect to draw a heart. Gamblers use
odds, which are, if anything, more confusing than probabilities. So we might say the odds of drawing a heart are
3 to 1 (often written as ¾) because you are three times as
likely to get one of the other suits as you are to get a heart.
A final way of representing a chance is to use a ratio of
percentages. You might say the chances of drawing a
heart are 25:75 in that you would get it right 25 times to
75 times you got it wrong, though this representation is
almost always used when there are equal chances of getting one thing or another, making it a 50:50 chance. This
is the same as a 1 in 2 chance, or a probability of ½, 0.5 or
50 per cent. In betting terms it is an evens chance.

Back with our roulette wheel and its 1 in 2, or 50:50
chance of winning or losing. This isn't a safe enough bet
for casinos, which as businesses want to make sure that
they will get a profit. So they add a zero slot (often there
are two of them on wheels in the US). If the ball ends
up in this slot, no one wins but the casino. It should be
crystal clear for players what this means – long term,
the casino *will* win. But of course this doesn't mean that
lucky players can't clean up – as long as they stop while
they are ahead.

A roulette wheel is a physical device, and as such is
not a perfect mechanism for producing a random number

between 1 and 37 (or 38 in the more money-grabbing US casinos). Although wheels are routinely tested, it is entirely possible for one to have a slight bias – and just occasionally this can result in a chance for players to make a bundle. It certainly did so for 19th-century British engineer Joseph Jagger, who has, probably incorrectly, been associated with the song 'The Man Who Broke the Bank at Monte Carlo', which came out around the same time as Jagger had a remarkable win in Monaco. The song probably referred instead to the conman Charles Wells, who won over a million francs at Monte Carlo and did indeed 'break the bank'. (This doesn't mean that he cleaned out the casino, simply that he used up all the chips available on a particular table.) There were many attempts to find how Wells was cheating, but he later admitted that it was purely a run of luck, combined with a large amount of cash enabling him to take the often effective but dangerous strategy of doubling the stake on every play until he won.

However, Jagger probably deserves the accolade more than Wells, as his win was down to the application of wits rather than luck – and was even more dramatic.*

---

* I have always found it bizarre that casinos consider it to be cheating if players use skill to win. Unless there is a fault with the wheel, there is no skill in roulette, but there certainly can be in games like blackjack, where there are a limited number of cards available to play, and so by counting the cards that have already been dealt, a player with a superb memory (or a concealed computer) can increase their chances of winning. Apparently you get thrown out of casinos if you are caught doing this and barred from re-entering. Imagine if athletes were banned from their sport if they showed any sign of skill. It just demonstrates what

Jagger finally amassed over 2 million francs – the equivalent of over £3 million (or US$5 million) today. He hired a number of men to frequent the casino and record the winning numbers on the six wheels. After studying the results he discovered that one wheel favoured nine of the numbers significantly over the rest. By sticking to these numbers he managed to beat the system until the casino realised it was just this wheel that was suffering large losses and rearranged the wheels overnight. Although Jagger soon tracked down the wheel, which had a distinctive scratch, the casino struck back by rearranging the numbers on the wheel each night, making his knowledge worthless.

In a casino, with the odd exception like Jagger, chance is under tight control, but when the second type of randomness, chaotic randomness, is in action, control rarely lasts long. Where classical randomness involves truly unpredictable individual events, chaotic randomness isn't actually random at all. Think, for example, of attempts to predict when an earthquake will occur or what the weather will be weeks ahead. The individual events in such chaotic systems *are* predictable, but in practice the interactions of the elements in the system are so complex that it only takes very small changes when chaotic randomness is in play to make massive differences in outcomes.

Where the random events of classical randomness come together to form a predictable distribution, chaos

---

should be obvious: casinos aren't a way of playing fair games, they are businesses designed to take money off people.

refuses to be so easily predicted. In chaos, individual items don't fit into neat distributions, meaning that chaos can bring with it huge surprises – what Nassim Nicholas Taleb refers to as 'Black Swans'. In a sense, chaotic randomness is far more random than true, classical randomness.

CHAPTER 2

More random than random

Both types of randomness – classical and chaotic – can catch us out, but there is nothing more dangerous than when we try to apply the rules of one kind of randomness to the other. The distributions of classical randomness work well in physics, or to describe how the heights of human beings vary. They describe perfectly the outcome of a fair casino game. But if we try to use these distributions to predict the behaviour of the kind of chaos that lies behind an earthquake, we will eventually come a cropper – and because of our pattern-seeking natures, we can easily rationalise a chaos moment away, pretending that it is just a blip we can ignore in an otherwise classical distribution, rather than the definitive action of a chaotic system.

Take an example derived from Bertrand Russell, used by Taleb in *The Black Swan*. Consider the life of a turkey. This is a particularly thoughtful turkey, which makes predictions about its future happiness and well-being. It looks back over its life so far and sees a normal distribution (of which more later) of good and bad days. On the whole, by assuming it is dealing with classical randomness, it can predict the range of its positive and negative experiences, and how the good and bad days will be distributed. And then Christmas comes. Chaos intervenes. There comes a point that is way off the scale as far as all its experience to date goes.

We are good at being like this turkey. Every business puts a huge amount of effort into putting together a budget and forecasting what its performance will be like over the next year. There are even painful post-mortems, examining why reality differed from the prediction. These businesses (and you can apply the same picture to economists and politicians) are turkeys, merrily predicting the future from the past and getting upset when chaos steps in. It's not your forecast that's wrong, guys – no need to have a post-mortem – it's the assumption that you can make effective predictions in most real-life circumstances.

To take on a more life-and-death example, think of a plane crash. This is another case of a chaotic intervention in the generally classical distribution of experiences of flight. Here we can see a particular danger that emerges from our inability to handle this kind of randomness. We think plane crashes are much more likely to happen than they really are. This is because we are presented with them much more often in the media than their impact deserves. As a result we are much more scared of travelling by plane than we are of going on a road.

Typically there are one or two thousand deaths in plane crashes worldwide each year (many of them in smaller airliners that most of us don't use). By comparison, around 1.25 *million* people are killed on the roads each year around the world. Yet a combination of media exposure and our difficulty handling randomness means we get particularly scared of flying. I put my hands up. I hate flying. It just *is* scary if you think about what could happen in a plane crash. Yet experiencing this is a very unlikely occurrence.

What's more, if a plane crash is reported on the news, your awareness of the dangers of flying goes up, just like the dangerous taxi driver effect in the previous chapter. There is no reason whatsoever for the risk to go up – but because we are more aware of the possibility, taking your next trip by plane feels more scary. A similar thing happened on a much larger scale after the 2001 attack on the World Trade Center in New York. In reality, because of the security clampdown, terrorist incidents were much less likely immediately after this event, yet everyone felt that they were at greater risk, thanks to the human inability to deal with randomness and chaos.

The success factor

One major implication of our difficulty with chaotic randomness is that we ascribe much to talent that, in activities where chaos has the upper hand, really should be allocated to luck. Success as an author, or investing in stocks and shares, or in running a large company is primarily down to how well the books, investments or companies fare in terms of what chaos can throw at them. The big spikes (the equivalent of the turkey's Christmas) far outweigh the subtle influence that having extra-special talent on board can bring to the party.

This doesn't mean the inverse – that it's easy to succeed with no talent – is true (despite the evidence of some TV celebrities). There is usually a competence threshold below which bad performance makes failure inevitable, but as long as you have a reasonable level of competence it is chaos that determines who will be labelled the super-talented and the big successes in businesses where chaotic randomness reigns. This isn't true of all business activities at all times of course. It's interesting to look at chain restaurants as an example that is, for the moment, relatively free from chaos.

If you run a McDonalds franchise you will probably exist in a comfortable existence of relatively small variations from prediction. Your budgets will be meaningful. While the exact numbers of customers will vary randomly, it will be within easily maintained limits, predicted by a handy distribution. In ordinary circumstances, demand for hamburgers has no good reason to be chaotic. Hamburger restaurants can fail, but it can usually be put down to clear causes, like having the wrong location. With an appropriate footfall and the right product at the right price, people will buy hamburgers. However, you shouldn't feel too safe. The life of turkeys is inevitably chaotic, but for the hamburger restaurant a major change in the environment can introduce chaos. If you have a major outbreak of salmonella, your business could disappear. Or the council could close the road you are based on, taking away your trade.

Longer term, even with such a solid business, the world could change, producing a dramatic move away from your products. Could anyone have imagined in the 1980s that Kodak would go into bankruptcy? Yet changes outside of the company's ability to predict took away its trade. Kodak was guilty of thinking that the chaotic intervention of digital photography was a blip that could be ignored. It thought it knew the photography business like no one else. Instead, the introduction of digital produced a transformation of the industry. Kodak moved too slowly and too late and suffered as a result.

The way that chaos creates apparent experts that in fact have no expertise can be demonstrated in the form of a classic scam. Let's say you were an unscrupulous person and wanted to make a lot of money from punters who like to gamble on the horses. You can offer them a prediction technique that seems to guarantee sure-fire success. You can sell people a scheme that genuinely enables you to predict the winners of a whole string of races correctly.

First, select a series of races in each of which there is only a small number of runners. For simplicity of doing the sums, I'm going to imagine each race has just four horses in it. Now advertise your amazing betting system. And wait for the punters to pay you a large amount of money for the output of randomness. It works like this. You will provide people with a prediction of winners – they pay you for the advance knowledge. To prove your system works, you will give them the first four winners for free. After that, they have to pay you £1,000 a time.

Let's say 4,096 people sign up for the first free prediction (you may well get many more to do this – it's free, after all – but it's a convenient number for the example). You simply split your punters into four groups, giving 1,028 customers the prediction that the first horse will win, the next 1,028 that the second horse is the winner and so on. At the end of this stage, you will have 1,028 punters who received a successful prediction. Discard the rest. Now repeat with your successful punters in the next race, dividing the 1,028 into four groups of 256 and predicting a different winning horse to each group.

Of the punters in the second race, 256 will be winners. Repeat this process again, tipping each of the four horses in the next race to 64 of your remaining punters. Then repeat again in the fourth race, dividing your remaining 64 punters into four groups of sixteen. Sixteen people will once again be correctly told the winner. These sixteen people will now have been given the winners to four races in a row. A fair number of them will have enough faith in your system to pay up for your next random prediction.

There are a number of ways to run the scam. You could, for instance, charge an increasing amount for each prediction, refunding the failures. But however the finances are organised, with a large enough set of punters at the end of the process, you will have a group of people who are convinced that you are an absolute genius. Because you will genuinely have perfectly predicted the winner in race after race. In reality, you had no skill, no talent whatsoever (except a talent for deception). The only reason you appeared to predict the answer is that the sixteen remaining punters were lucky enough (in this case, just a 1 in 256 chance) to be in the winning group each time.

The same type of random allocation to success applies to pretty well every stand-out phenomenon in publishing or investment. It's not that those involved have any special talent, far above the common herd, just that there are thousands of others who pass the basic competence test but who are the equivalent of one of the punters with a failed sequence of predictions. The lucky ones succeed dramatically – the majority are unlucky and don't achieve that outstanding success. The only difference between the real world and the scam is that in the real world there is no scammer. These processes with chaos at their heart don't need the help of a conman: they are a natural consequence of having large numbers of parallel activities in a social setting.

Random success

Because of our need for patterns and to have clear causality, we find it very difficult to accept that there isn't something special about J.K. Rowling or Bill Gates or Richard Branson or George Soros, so we attribute some kind of wonderful skill to them. But they don't need such a skill any more than the winning punters do to rake in the money. Of course, it's very unlikely that any particular individual happens to be the one that receives the boost from chaos. The vast majority of their peers will be like the punters in the scam sequences that don't deliver. But some will inevitably succeed this way.

In some ways it's like a lottery. The chances of any individual winning are millions to one against – but the individual with the right ticket will be the winner. There is no skill involved. No expertise. Just luck. You have to have the basic level of competence to remember to buy the ticket before the draw takes place, but after that, how well you do provides no reflection on the quality of your performance.

The big difference between the lottery and publishing or stock trading is that while we can't predict what the winning number will be, we *can* predict that there will be a winning number drawn in any lottery draw. The actual drawing mechanism is a process of classical randomness rather than chaos. By comparison, we can't predict that a book by the next J.K. Rowling will be published next Saturday, or even this year. All we can do is say that there is likely to be another at some time. But the comparison with the lottery is still useful in emphasising how chaos fools us into thinking that there is expertise at work where in truth there is randomness.

It has been suggested that a lottery isn't a good parallel for bestselling authors and wildly successful business people and investors because there is a fundamental difference. The lottery obeys known probabilities – we know what the chance of winning with a particular ticket is – and the amount of payout on a lottery is known and limited, where a chaotic random event is so far outside expectations that it is almost limitless. This is a futile argument. The importance is not the mathematical values but the impact on human beings. A major lottery win is so unexpected and so transformative that the impact on the individual is indistinguishable from a chaotic random event.

You may feel that my apparent attack on J.K. Rowling, suggesting that her success is down to luck rather than being a great writer, reflects the bitterness of an author who hasn't achieved that same level of success, but it really isn't. Let's look at three relatively recent huge successes in the literary world – *Harry Potter, Twilight* and *50 Shades of Grey*. Why did they succeed in such a vast, till-shattering way, when so many other books didn't? What can we say about these books?

We certainly can't put their success down to being great literature. The *Harry Potter* series was competently written, though arguably was not better than average, and there were certainly plenty of competitors with the same potential to appeal to adults and children that were technically better pieces of writing. It would be hard to find a literary expert who wouldn't argue that the *Twilight* series is markedly worse in quality of writing than *Harry Potter*, while few would call *50 Shades* anything other than a literary disaster. This isn't about quality of writing.

Nor can we say that these books were great successes because they were wonderfully original. Each had clear precursors. And though fans might point to characters we can support, or clever plot twists, or the power of word of mouth, the reality is that in every case there will be many better examples that have not achieved blockbusting success.

The overwhelming success of these books is quite simply the work of chaos – randomness that we could never predict nor seek out. It is utterly pointless for a publisher to look for the 'next *Harry Potter*' or 'next *Twilight*'. The very action of doing so implies that someone can see a pattern where none exists. Anyone who does so is hunting a chimera. All publishers can really do is to ensure that the competency threshold is passed and get the books out in front of an audience to give chaos full and magnificent rein.

I need to reiterate that I'm not saying there is no benefit in writers (or investors or entrepreneurs) putting in lots of nose-to-the-grindstone hard work, or in honing their skills, or researching their topic well. The process is not *all* down to luck by any means. This groundwork will help make success possible. It buys you the ticket for the lottery. However, being skilled at the job only ensures an entry to the middle ground. To shoot into the stratosphere takes that push that can only come from chaos. Similarly, this doesn't mean there is no point in promoting a book. Good visibility can easily push a book into a bestseller list and give it some temporary momentum. This is clear from the W.H. Smith experiment.

A contact who had been a senior manager at newsagent and booksellers W.H. Smith once told me that, as an experiment, they took a book that was hardly selling at all and gave it the full, front of store, big display exposure. It shot into the top ten of books in the UK for that week, and did well for weeks afterwards. But all such action does is to add a few thousand extra sales and make the author and the publisher a comfortable little profit. If I mentioned the book's title it would mean nothing to you now. Promoting a title doesn't produce the chaos-fuelled blockbusting bestseller. That's about millions of sales, not an extra few thousand.

How we hate this. It's not fair. It really isn't. Getting a chaos success is not a reward for hard work or excellence. It's luck. Fairness doesn't enter into it. Because we hate it so much, and because we are so desperate to see patterns we will always be able to dream up a reason to explain what happened. It was the right time for a book like this. The author had incredible hidden talents. She tapped into the zeitgeist. She had the right platform. But it is so much justification after the fact. Just listen to the news to see this in action on a much wider scale. Every time the stock market moves up or down, the analysts will tell you it's because of this and that. They may come up with some plausible reason – an announcement from the central bank or a company's latest results. But it's a convenient fiction. They cannot truly know what is happening. The chances are that it was chaos at work.

Scarily, the same goes for politics. Whenever an election comes around there will be much focus on the current administration's record in office. Perhaps, for instance, on their watch the country suffered a financial disaster. Do the voters really want more of the same? Yet it's a political fiction that a government or a president has any control over the economy. It's far too complicated and chaotic a system. Politicians will always claim that it was their policies that caused good things to happen, and circumstances beyond their control that caused the bad things. In reality, hardly anything on the scale of developments in the world's economy can be given a clear causal link to the policies of politicians.

Superstition conjures causation

This urge to find causes at any cost, to satisfy our pattern-driven brains with made-up explanations, has a name. It's called superstition. Whenever we assign a cause to an occurrence that isn't really connected to it, we are acting superstitiously. And it's a very natural thing to do. In experiments with pigeons it was accidentally discovered that if the pigeons happened to perform a particular movement a few times before they were fed, they would start performing that movement whenever they were hungry. They had become superstitious. They had seen a pattern that linked the movement to being fed and thought that the movement *caused* the food to arrive.

My dog seems to have developed the same kind of ritual superstition. Every morning, before she is fed, she expects to be stroked first on her stomach, then on her ears. Once that has been done she will go to her bowl, ready for her food. But she actually puts off getting her breakfast to go through the ritual – that's the strength of superstition.

We are used to superstition as a silly belief that supposedly provides a causal link between, say, walking under a ladder and something bad happening. Or a link between the discovery of a 'witch' in the village and cattle falling ill or crops failing. Most of us would laugh at this kind of superstition, finding it hard to believe anyone could fall for it. Yet many people today believe that taking homeopathic pills makes them better, or that being in the proximity of phone masts make them feel ill. These are very much the modern equivalent of the village witch story. What we've got is correlation without causality. Just because two things happen in a similar location or at a similar time doesn't mean that one causes the other. Superstition turns a correlation (events that happen at a similar time or place) into a false cause.

We may spot superstition at work in the above examples, but it is much more common than we think. It happens whenever someone says 'X happens because of Y' but there is no evidence that one causes the other. This can happen by coincidence, or where Y actually happens because of X – or where there is a third cause for both X and Y. So, for example, for several years after the Second World War in the UK, the rate of pregnancies was correlated with the import of bananas. In years when there were more bananas imported, there were more pregnancies. Fewer bananas, fewer pregnancies. But the bananas did not cause the pregnancies – both were probably the result of a third cause, such as changes in economic circumstances.

The typical 'X happens because of Y' claims we hear all the time are less outrageous than bananas causing pregnancies, but they still need the same examination to see if there really is a causal link. For many years in the UK there has been evidence that girls who attend single-sex schools perform better academically. There have been all manner of mechanisms put forward to explain why attending one of these schools would improve an individual's performance, often including the fairly obvious lack of distraction by boys. However, there is a big assumption being made here.

We don't know that being at a single-sex school *is* the reason these female students do better than their contemporaries elsewhere. There are other ways that these schools differ from a typical school. Many single-sex establishments are private. Many private schools have smaller class sizes than state schools – and smaller class sizes may well contribute to better results. Students at private schools will tend to have richer parents, which contributes a number of factors that could improve exam results, from the ability to bring in tutors to a different social attitude to exams and the expectation of going on to higher education. For that matter, proportionally far more single-sex schools than schools overall are selective. Such schools only take students with a minimum level of academic ability, producing a self-fulfilling prophecy. It is flawed thinking indeed to make the immediate leap into assuming that the single-sex nature of a school is the *cause* of students getting better results.

The need to be sure of a true causal link is particularly important when we are trying to explain something that varies from day to day in a natural way. To attribute any cause, we need to be able to find a significant change on top of this background noise. Imagine for a moment that the stock market went up or down totally randomly. Let's say each share value was allocated that morning by a random number generator. The fact is that today, as a result of those random changes, the market would either have risen or fallen from yesterday. That is inevitable, even despite there being no underlying cause whatsoever. And immediately analysts would find a reason for it.

Whenever a commentator says 'The stock markets are down because of the announcement by the German Chancellor' or 'This country's GDP went up because of our party's policies' or 'The value of houses in our neighbourhood have gone down because these types of people are buying them' or 'My football team is doing better than last year because of our new manager', take a moment to ask 'How do you know this?'; 'What evidence is there to show that one thing causes the other?' Correlation is not enough.

It isn't just the person in the street who can confuse correlation with causality. In 2004 a Swedish scientist called Jarl Flensmark published an academic paper titled *Is There an Association Between the Use of Heeled Footwear and Schizophrenia?* What is disturbing is that despite apparently asking a question in the title of the paper, he presents the hypothesis in the text as if it were a statement of fact: 'Heeled footwear began to be used more than 1,000 years ago and led to the occurrence of the first cases of schizophrenia.' Flensmark then shows a parallel between the growth of heeled shoe production and an increase in the prevalence of the disease.

We are told that the first known examples of heeled shoes were in Mesopotamia, as were the first institutions for mental disorders. A whole string of European royals are listed as possible victims of schizophrenia and who were also known – or at the very least thought – to have worn heeled shoes. Flensmark notes that it is the upper classes around the world that typically wore heeled shoes first – and it is the upper classes who were more likely to report symptoms that would now make doctors suspect the existence of schizophrenia. The pattern, Flensmark suggests, is simple. After heeled shoes are introduced, the first cases of schizophrenia appear, and as wearing of the shoes grows more popular, so do the frequency of attacks of the disorder. Simple cause and effect.

Flensmark comes up with an ingenious, if rather intricate explanation for why walking in such shoes could have an influence on the brain. But there are so many opportunities here to confuse correlation and causality. Heeled shoes have, as he suggests, typically first been taken up by the upper classes, because they are impractical, and the appeal of impracticality usually only develops once you don't have to worry about where your next mouthful is coming from. Wearing such shoes also will tend to increase as society as a whole gets wealthier and more sophisticated. Yet the trappings of class, wealth and sophistication are also more likely to result in more *reporting* of illness, mental or otherwise. If life is a constant struggle, you either die or you get on with existence despite any illness. In a primitive society like medieval Europe, there is no medical safety net. Being seriously ill and staying alive is a luxury only available to those who can afford it.

What seems to be recorded here are two separate causal links, which when combined result in an unrelated correlation. It seems entirely reasonable that wealth and being of a higher class cause the increased wearing of heeled shoes. And it also seems likely that wealth and being of a higher class produced increased reporting of the symptoms we associate with schizophrenia. But there is no reason to deduce that the shoes caused the mental illness. In fact if there were a causal link, the more obvious one might be that schizophrenia caused sufferers to be more likely to wear heeled shoes, which are hardly a rational piece of footwear. There are, no doubt, many other potential causal structures as well, but the point is that an academic has made an assertion of causality despite there being absolutely no real grounds for making it. Humans – even academics – need their patterns.

A natural cycle

Not only can we get a cause wrong, we can easily attribute cause to what is simply the random nature of a sequence of events that naturally go through peaks and troughs without there being any great cause for any particular value. Just think of sitting by a beach, watching the waves. Sometimes they are big, sometimes they are small. Sometimes they crash to the shore, sometimes they ripple in with hardly any strength. We don't say, 'Oh, that wave was big because a seagull flew past,' we just accept that this is the nature of waves. When we look at the daily variations up and down in the stock market, or the behaviour of a football team from match to match, the chances are that we are observing similar waves of natural 'up and down' nature in action. There really doesn't have to be a cause – but we are desperate to find one.

What is happening with the waves – and with many of the upward and downward twitches of the stock market – is time-based clustering. Clustering is one of the most misleading aspects of randomness. When things are distributed randomly – in time or space – you get periods or points when the occurrences cluster together and you get gaps. This is just what randomness is like. A good picture is to imagine dropping the contents of a box of ball bearings on the floor. You would be very suspicious if they ended up in an identically spaced grid, each separated the same distance from the others. There would have to be magnets under the floor or some other cause to make this happen. What you expect is that some will be clumped together and some will be spread out with gaps.

We can accept this view of randomness for ball bearings, just as we can with waves. But as soon as we move to something that has an effect on people, we forget that randomness is like this and start looking for patterns. The spatial equivalent of looking for reasons for movement of the stock market is looking for causes of clusters of events that occur physically near together. When something goes wrong in a cluster we want to find a scapegoat. Traditionally we would have looked for a witch – these days we look for a phone mast or a nuclear power plant or some other local cause. Of course some clusters do have causes, but the great majority, including the vast majority of cancer clusters that are investigated, turn out to be down to the simple nature of randomness.

Clusters of events are one thing, but perhaps the hardest type of randomness to accept of all is that involved in being a huge success. As we have seen, we rebel at the thought that a superstar only achieved great things as a result of luck. To get a better handle on this, it can help to have some simple tools for the job.

CHAPTER 3

A measure for luck

I love reading books about successful people. It is very entertaining to discover how Bill Gates or Steve Jobs or Richard Branson or Jeff Bezos achieved their amazing success. And I'm not alone. These are superb stories. But be wary if the business biography doesn't stress the importance of luck. (Gates, for example, had his big break because his main rival decided to take his plane for a flight, rather than attend a meeting with IBM.) If that business book tries to tell you how following the pattern of these hugely successful people can make you hugely successful too, the author is indulging in superstition, or trying to rip you off. You can't learn how to be lucky. You can learn how to make the best of what luck throws at you, but that is a way to reach the middle ground, not to become fantastically wealthy. Enjoy the stories in superstar business biographies, but don't be taken in by correlation mistaken for causality.

One interesting way to get a feel for how much a particular type of success is driven by chance is to use a technique devised by the great Richard Feynman to help understand quantum physics. Feynman was a huge character and is still a hero to many physicists. Seeing one of his filmed physics lectures can be surprisingly entertaining, helped by the fact that his strong New York accent means that it is a little like being taught science by Tony Curtis.

Feynman sprang to fame outside his field when he was on the commission investigating the Challenger shuttle disaster. He discovered a possible reason for the accident, and rather than attempt to describe what happened in the fusty wording of an official report, in front of the cameras he dunked an O-ring* into his glass of iced water and demonstrated how it lost some of its flexibility. But in physics, his greatest contribution, for which he won the Nobel Prize, was in quantum electrodynamics.

This is the science of how matter and light interact, something that is happening all the time, whether it's photons of light hitting your retina and enabling you to see, or light hitting the Earth from the Sun and giving it the heat that made life possible. Feynman took an approach called the 'sum over paths'. The idea was simple but startling. Imagine you have a beam of light coming from a light source, like the beam from a laser pointer that reflects off a mirror. High school science tells you that the light travels towards the mirror in a straight line, bounces off at the same angle at which it had arrived, and travels off, once more in a straight line.

Feynman's technique, based on the strange behaviour of quantum particles, says that instead, let's look at what happens if the photon of light travelled along every single one of the infinite possible routes that it could take between its source and its destination. All routes would not have the same probability of being used, but all would have the potential to be travelled.

---

* An O-ring is a circular rubber seal of the kind that failed in the Challenger disaster.

Now add in one final aspect. A photon of light has a property called its 'phase' that varies with time as it moves along. It's a bit like it carries a little clock, and the phase shows the direction of the second hand of that clock (except the clock hand rotates a whole lot faster than a typical timepiece). When we add together all the possible paths, those that end up with the phase (clock hand) pointing in the same direction add together and reinforce each other, while those with phase in opposite directions cancel each other out.

If you add up all the infinite possible paths with their varying probabilities, using the phases, what comes out is what we actually observe – the photon appears to come in on a straight line, reflect at the same angle as it came in and go off in a straight line. If that was all that the method did, it would be a waste of time and effort – but in practice it allows us to predict why all kinds of strange optical phenomena happen. For example, when a CD reflects light and we see little rainbows, what's happening is that different energies of photon are travelling along different summed paths – and the reflection takes place at an angle that is totally different to the angle of the incoming light. In this case, unlike a traditional mirror reflection, the sum over the paths predicts something other than a straightforward reflection – and that's what happens.

We can apply a similar approach of using the sum over all paths in a loose fashion to explore the way different successes or disasters are related to randomness. Take the example of a lottery winner. There is no skill involved: all tickets have the same 'ability'. When there's a fuss on the news, we see a specific path that the winner took, which led to success and riches. But if we take a look at the sum over all paths of people who bought tickets for the same lottery, we get a different picture. Overwhelmingly the action led to them losing their money. The composite lottery player summed over all the paths is a loser.

Let's try a similar approach with someone who writes a novel. Talk to the general public about being an author and they immediately think of the J.K. Rowlings of this world, selling millions of books and making vast quantities of money. But consider the sum over the career paths of all novelists with the basic competence and the outcome is pretty miserable. The average published novelist sells around 1,000 copies of his or her book. They will be lucky to earn £1 per copy. Such small numbers dominate the sum over paths. Contrast this, for example, with the sum over career paths of plumbers with a decent competence. As individuals they don't get hugely rich like the lottery winner or the super-successful novelist. But the sum over all paths for plumbers will be much higher than it is for novelists.

The inhuman economist

One group of people who ought to appreciate the whole sum over paths model is economists, yet economists rarely understand randomness and its impact. Economists appear to make predictions that defy all the chaos that is out there – but their success is an illusion. The illusory nature is clear from the frequency with which economists contradict each other. It is the only discipline that calls itself a science where year after year theoreticians with diametrically opposed ideas win Nobel prizes. That's almost a definition of an art rather than a science.

One cause of their errors is that in order to predict human behaviour, economists assume that individuals will behave 'rationally', which to economists means acting in such a way as to maximise financial benefit. The problem with this straw man picture, often called 'homo economicus' is that it takes a very narrow-minded approach to how human beings really behave. Not only do we humans often fail to act rationally, but even when we are apparently rational, the human assessment of maximising benefit is rarely purely about optimising financial gain.

Honest economists, who haven't fooled themselves into believing their own hype, will admit that their actions are like the example sometimes used comparing the scientific approach to finding lost keys on a dark street at night. The story goes that you meet a scientist who has lost her keys and join her in searching for them. She is looking under the only streetlight in this road. After searching for about fifteen minutes, you ask, 'Are you sure this is where you lost your keys?' The scientist stops for a moment. 'No,' she says, 'I think I lost them in the next street. But there are no streetlights there at all.'

The argument is that while at first glance the scientist seems to be stupid, there really is no point looking where there is no light. However poor the match to what's likely to be true, light is necessary to see something.

I personally think this is a poor example of science, as a good scientist would use other means, like feeling with their foot in the correct street, but that's a different story. Economists, in assuming 'rational' behaviour, are often simply doing the best they can with no other information – but we need to be aware that they are almost certainly not achieving a worthwhile picture of reality.

A very useful demonstration of just how far 'rational' economics takes us from reality is the ultimatum game. It's a very simple, and extremely informative test. In it, you and a second person are asked to make a decision about some money. The two of you mustn't discuss your decision in any way. You are given £1 (say) to share. There are no strings attached, it is a genuine gift, the pair of you simply have a decision to make before the money is given to you.

First, the other person decides how the money is to be split between you. They can split it however they like. The money can be split 50:50, they can keep all the money to themselves, they can give you a penny and keep the rest … or split it any other way they like between the two of you. You then say either 'Yes' and the two of you will get the money, split the way the other person decided, or 'No' in which case neither of you gets any money. There can be no discussion between the two of you.

This game has been undertaken many times over the years in many circumstances. (Who says economists and psychologists don't know how to have fun?) The economically rational thing for the person in your place to do is to say 'Yes' as long as the first person gives you something. Anything. Even if you're only offered a penny, it's money for nothing. Why would you logically say no? (The traditional economist's picture of a human being wears a wristband inscribed 'WWMSD?', standing for *What Would Mister Spock Do*?) In practice, though, a real person in your place tends to say 'No' unless they get what they regard as a fair proportion of the money.

What counts as a fair proportion will vary significantly from culture to culture. Some will accept as low as a 15 per cent cut, others expect a full 50 per cent – but in Europe and the US we tend to expect around 30 per cent or more before we will say 'Yes'. Anything less and we feel hard done by, and are prepared to take revenge, even if it has a financial cost.

What the experiment shows is that we consider trust and fairness worth paying for. We are willing to lose money in exchange for putting things right. If human logic were based purely on economics, then this just doesn't make sense. You always should take the money if any is offered. But your brain makes decisions based on a much more complex mix of factors than finance alone.

This is not to say that finance doesn't have a significant input into decision-making – and any psychologist who expects a Western player always to demand 30 per cent or more really hasn't thought through the real-world version of this game. If, for example, a billionaire decided to play this game, and offered a total stake of £10 million, the chances are that you would accept being offered just 1 per cent: £100,000. Unless you are extremely rich yourself, that is just too life-changing an amount of cash to turn down in order to teach someone a lesson and punish their lack of fairness. In reality, you would swallow your pride, ignore the psychologists and take the 1 per cent cut.

It's an interesting exercise to think to yourself just how little you *would* accept in such circumstances. Where between £100,000 and £1 (which most people would reject) would you draw the line on an ultimatum game with £10 million on offer? I think I might cave in for as little as £100, which would be just 0.001 per cent of the total amount – but then I'm cheap. That rough 30 per cent break point very much depends on the amount of money at stake.

Gambling with chance

It's almost inevitable that one of the ways we get a better understanding of our attitude to economics is through a game. The urge to place money on the outcome of an event, whether classically random or apparently based on skill, is one that goes back as far as most other markers of civilisation. Even the kind of people who don't regularly buy lottery tickets or gamble on the horses or ever walk into a casino may occasionally rise to a bet – a financial stake in what is hopefully, but in truth is rarely, a calculated risk.

As we have seen, not all randomness behaves according to a neat distribution, but many of the standard tools of gambling, from the throw of the dice to the chances of getting a particular poker hand, do. The oldest known evidence for gambling comes from knucklebones or astragali which have turned up in archaeological digs dating back thousands of years. These playing pieces are chopped from the end of an animal bone, shaped to be four-sided and so making a form of die that can be bet on. Sophistication was later added by combining six-sided dice with a board and counters in tables (the ancestor of backgammon) and by the introduction of playing cards – both these developments allowed a combination of skill and chance to make the betting process more interesting.

However, the development of such games did not take away from the simple attraction of betting on the outcome of classical randomness, whether it is a simple coin toss, dice throw or card draw, or the more sophisticated mechanisms employed by the casino. All life involves taking risks: there is something appealing about this controlled form of risk that limits chaos and allows us to dream of easy money.

Something that gambling uncovers neatly for us is the relationship between chance and time. The risk that randomness holds over our heads dissolves with time. Before the throw of the die we have a one in six chance of winning. After the throw we either have a 100 per cent chance of having won or a 100 per cent chance of having lost. The risk has disappeared with the outcome. This is one of the reasons that the concept of a time machine is so alluring. With a time machine that allows travel back from the future we can pop ahead, check the winning ticket in a lottery and return to make a purchase without any risk.

The idea of making mathematical predictions about the future, of using numerical analyses of probability to quantify chance, seems not to have made a lot of impact until surprisingly late in the development of mathematics. The Ancient Greeks lacked the mathematical symbology to make a good attempt, though their lack of interest may also have been down to a kind of fatalism, an idea that the universe was in the grip of a chaotic randomness that was beyond human comprehension.

Even the innovative medieval Arab mathematicians don't appear to have got far with probability. It might simply have been that, up to the Renaissance, the lives of human beings seemed so much in the hands of fate, with practically everyone lacking control over their environment, that probability didn't even arise as a concept. Some have also argued that an unthinking following of religion tended to suppress any thought of probability. If every aspect of your future life is down to the whims of a god or gods, then there is little point in trying to predict what will happen.

The first real exploration of probability came in a book called *Liber de Ludo Aleae* ('Book of Games of Chance') written by the Milanese physician and obsessive gambler Girolamo Cardano. Cardano was born around 1500 and first wrote the book in his twenties, but did not produce the final version until 1565, just six years before his death. Even after he was in his grave it took another 92 years before his masterpiece made it into print.

It was Cardano who introduced the idea of representing the chance of something happening as a fraction. If I toss a (fair) coin, then the chance of getting a head or a tail is pretty well equal. It isn't actually equal on any particular coin toss as there is a small dependence on the way up you hold the coin before flipping it. If, for example, you start with the coin heads up, it will come down on heads around 51 cases out of 100. But let's assume that there is a straight 50:50 chance – an expectation of around 50 heads and 50 tails in 100 tosses – to understand Cardano's innovation.

He said that, because we could expect, on average, one toss in every two to result in a head, we can say the chance of getting a head (Cardano didn't use the term probability) would be ½. Another way of looking at it is that you get heads half of the time and tails half of the time. If you extend that to the chance of getting a particular value of card drawing at random from a normal card pack, then you would say the chance of getting (for instance) a Jack is ⅟₁₃ – because one in every thirteen cards is a Jack and you would expect to get a Jack one-thirteenth of the time. This is very straightforward stuff to us now, but it was a remarkable insight in Cardano's day.

What was particularly powerful about Cardano's assessment was that he didn't just work out the probability of, say, getting a six with a throw of the die (⅙), but also of getting a five *or* a six which is twice as likely, as two options out of the six match our criteria (²⁄₆ or, more simply, ⅓). For that matter, Cardano also explored the chance of getting a six with either of two dice. This is a considerably more subtle problem. A simplistic approach might be to say that if the chance of getting a six with one die was ⅙, then with two dice it should be ²⁄₆ or ⅓. The trouble with this is that if you had six dice, according to this reasoning you would have a ⁶⁄₆ chance – absolute certainty – of getting a six, where in reality it's perfectly possible to throw as many dice as you like and have none of them a six. Even more bizarre, if this were true, then if you threw seven dice, the chance of getting a six should be $\frac{7}{6}$ – it would be more than certain. Yet getting more than 100 per cent of something isn't possible outside of a reality TV show like *The X Factor*, where we are regularly told that contestants give 110 per cent.

What Cardano realised is that you have to go about the combination of probabilities rather sneakily. If the chance of getting a six with your first throw is $\frac{1}{6}$, then the chance of *not* getting a six is $\frac{5}{6}$. Similarly, the chance of not getting a six with the second throw is also $\frac{5}{6}$. The chance of not getting a six in either case is $\frac{5}{6} \times \frac{5}{6} - \frac{25}{36}$. So the chance of getting a six with either is $\frac{11}{36}$ – slightly less than the simple result of doubling the probability for one die.

Other mathematicians, notably the French polymath Blaise Pascal, extended Cardano's thinking to produce a theory of probability that made it much easier to make predictions of classical randomness. Cardano, for example, had failed to solve a significantly older problem that dates back at least to the 15th century. Imagine you are playing a game where the winner is the first person to reach ten points. Because of circumstances beyond your control, the game has to be stopped when one player has eight points and the other five. How should you divide the winnings, assuming that there was some kind of financial stake in the game? After all, either player could have won, but the player with eight points seems more likely to win, so should get some financial advantage.

Pascal came up with an appropriate mathematical solution. Note, by the way, that there is no way he could forecast what the actual outcome could be. He had to make the assumption – and it's a huge assumption – that the better player would win eight games out of thirteen where the other player would only win five out of thirteen. In reality all kinds of things could happen. The person who was ahead could tire more quickly and lose many more of their later games, for example. Pascal's calculation of probability could not predict the future (something we should always remember) and it is very likely that it would not match the actual outcome had the players continued to the end of the game. But dividing the winnings 8:5 was still the best guess given the information available, and so better than nothing.

With the work of Cardano and Pascal, it was becoming increasingly possible to use probability to make predictions about some controlled kinds of randomness. If probability is one face of the way we can try to use numbers to tell us more about what is going to happen in the future, the other face is statistics.

CHAPTER 4

It's all in the stats

The term 'statistics' comes from the same origin as the word 'state' – statistics were originally pieces of information about a nation or state. In the modern sense, the aim of statistics is to take information about a sample of individuals and to make deductions about their future or about a larger whole. Where individual components of a system operate in certain predictable ways (something that is often not true of people), then the techniques of statistics enable us to get a feel for the behaviour of the overall mass even though we can't say anything much about individuals. As we will see later, this is particularly useful when trying to understand the behaviour of something like a gas containing vast quantities of molecules. We can't predict how each individual molecule will go about its business, but we can make very good predictions as to how the gas as a whole will behave.

The man who started us in the direction of using statistics beyond simply collecting data was not a high-flying mathematician or scientist, but a seller of buttons, John Graunt, who despite considerable prejudice against his trade became one of the early members of the Royal Society. (In the early days, the Society had a much more interesting mix of characters than it does today.) Graunt published a groundbreaking book called, in the wordy manner of the time, *Natural and Political Observations Made Upon the Bills of Mortality*, making use of information on births and deaths in London from 1604 to 1661.

What is particularly impressive about Graunt's work is not just that he assembled figures in a way that had never been done before, attempting for instance to show how deaths from plague varied from year to year; he also went beyond the numbers he had, making a first use of statistics to attempt to infer other values that weren't actually present in the data. For example, he tried to make estimates of the population of London, then a significant unknown, by making assumptions about how the number of births related to the number of potentially childbearing mothers. He even tried to estimate what percentages of a cohort of people (a group of people all born at the same time) would reach various ages before dying.

This analysis by Graunt, and work a few years later by the astronomer Edmund Halley, who took similar tables (this time for a German town) and calculated life expectancy at different ages, provided the groundwork for the insurance industry, a business that grew up in the coffee houses of London and spread to the world. In effect, insurance is a form of betting on risk to reduce its impact – attempting to tame randomness by spreading the risk. Most of the time, insurance is a business that benefits from the predictable distributions of classical randomness, but occasionally – when, for instance, there are major natural disasters – it is swamped by the reality of chaotic randomness.

What's it worth?

You might think that the realisation that we can't rely on the simple mechanics of economics when dealing with statistics about real people is quite a recent phenomenon – and it seems to be, as far as economists and stock traders are concerned. But some mathematicians have been aware of this for a long time. Back in the 18th century the brilliant German mathematician Daniel Bernoulli wrote a paper that showed how it was necessary to go beyond the apparently basic mathematics of chance to what he called utility: an attempt to quantify what something is really worth to you. And Bernoulli's big insight was that this 'utility' was not a constant across all human beings. It varied depending on your current position and your values, just as the decision of when to say 'No' in my £10 million ultimatum game will be hugely different for someone with no savings or income and a millionaire.

Another very valuable observation in Bernoulli's paper was suggested to Daniel by his equally gifted cousin Nicolaus. It shows that we have to be very careful about how we apply another probabilistic tool of the economist: the idea of expected value. This can be a very useful guide in some kinds of chance – but not, as the Bernoullis showed, in others.

The concept of expected value says that when we have different possible future gains, each with its own probability, we can compare the benefits by multiplying the gain by the probability. This sounds more complicated than it is! Imagine I have a choice between two investments. One has a 1 in 2 chance of gaining me £100 – if I lose out, I get nothing. The other has a 1 in 4 chance of my getting £200 – again, if I lose out, I get nothing. To find out which to choose, we multiply the expected gain by the chance of getting it. So in the first case the expected gain is £100 × ½ (because it's a 50:50 chance – I would win half the time), or £50. In the second case it's £200 × ¼ – again £50. Both options have the same 'expected value' of £50, so they should both be equally attractive.

If the second case had been a one in four chance of getting £160 – still a larger figure than the £100 in the first case – then the expected value is £160 × ¼ – just £40. So it is less attractive to go for the £160 than it is to stick with the £100. So far, so good. You can also combine different possible outcomes. If, for example, you have a 1 in 2 chance of getting £100 *and* a 1 in 4 chance of getting £200 from the same investment, the expected value is (£100 × ½) + (£200 × ¼) – your expected value is £100. But this reasoning breaks down in all too many real circumstances. The Bernoullis' example was an extreme one, but it highlights the point.

They imagined a simple mechanism for calculating the rewards from an investment. We toss a coin repeatedly until we get a head. If we get a head on the first throw, then the return is £1. If we get a tail we carry on. If there's a head on the second throw, the return is £2. If we have to wait until the third throw for a head we get £4. And so on. Now the question is, how much would you invest to get this return? The mechanically thinking, unimaginative economist would haul out his expected value calculator and would get quite a shock.

The chance of getting £1 is ½ – half of the time you would get a head on the first throw. The chance of getting £2 is ¼ – because you multiply the chance of getting to the second round (½) by the chance of getting a head on the second throw (½) to get ¼. Similarly, the chance of getting £4 is ⅛, the chance of getting £8 is ¹⁄₁₆ and so on. To calculate the expected value of the game overall, we multiply the different outcomes by their probabilities and add them together.

So we get $(1 \times \frac{1}{2}) + (2 \times \frac{1}{4}) + (4 \times \frac{1}{8}) + (8 \times \frac{1}{16})$ ...

There's a pretty obvious pattern. Every item in that series works out as ½. So the total expected value is ½ + ½ + ½ + ½ ... all the way to infinity. And the sum of an infinite set of halves is itself infinity. The expected value of taking on this investment (or wager) is *infinite*. What that seems to imply is that you should be prepared to invest anything, as much as is asked, in such a wager, because the expected return is by definition bigger than your investment. But thinking realistically, would you risk £1 million, say, with a 50 per cent chance of losing all but £1? Expected value is not an effective guide when the impact of losing the wager is catastrophic.

This paradox demonstrates, in many ways, the lure of the stock market. If you had stocks that you imagined, in the long term, would continue to grow and grow in value at a fast enough overall rate, then it really doesn't matter what you pay for them. You will benefit in the end. Unfortunately, this approach forgets two aspects of the chaotic nature of the randomness of the stock market – that markets crash and that businesses lose their reason for existence, leaving their shareholders high and dry.

I have personal experience of one of these happening. I don't often dabble in the stock market, but just before the banking crisis of 2008 I noticed that several of the bank share prices took a major fall, then recovered within a day. If you had invested in those stocks at the bottom of the fall you would have made 200 to 300 per cent profit on the day. That's quite a rate of interest. A couple of days later another bank's shares did exactly the same thing. It looked like easy money. So when the shares were about ⅒th of their previous value, I bought £200 worth. But this bank was called Bradford and Bingley. The shares continued going down and down – until they were suspended. The bank went bust. Its customers were protected by the government, but shareholders, including opportunists like me, lost everything. You may say I deserved everything I got, but it shows the danger of forgetting that chaos is always ready to take a hand.

This was an example of the impact of a financial crash, but no crash is needed for chaos to disrupt the expected value of shares, even shares in the 'solid' growth stock companies that have been so popular with investment managers over the years. These are companies like Coca-Cola and McDonalds, which seemed, like the Bernoullis' game, to promise pretty well infinite expected value. Investors have often flocked to them, yet in practice these shares have done less well than a broader portfolio like the FTSE 100 or S&P 500. More worrying still, though we can't predict *which* will suffer, we can say with some confidence that some of those 'solid as a rock' companies will end up in bankruptcy with valueless stocks.

Just take three examples. Pan Am was one of the absolutely solid airlines. It was a big name, big visibility company, the US flag carrier. Stanley Kubrick even envisaged Pan Am shuttles carrying passengers to his space station in *2001: A Space Odyssey*. Pan Am has gone. Less well known, Wang was a massive technology company in office computing systems in the 1970s. By comparison with Wang's integrated office functionality, IBM's new PCs seemed feeble. But within six years of the PC becoming widely available, Wang, which never really understood this technology, was bankrupt. And if you really want to see a giant fall, how about the one I've already mentioned, Kodak? A massive company that was safe as houses. A company that had already thrived for over 100 years. Until digital photography took most of its market away and it ended up in 2012 scrabbling for survival.

When we look at the Bernoullis' paradox – why we don't offer practically anything to buy into that investment with an infinite expected value – what we see is that we value certainty over risk at the hands of randomness. Say you invested £100. Fifty per cent of the time you would only get £1 back. The first return that would end up with you in profit is getting £128 on the eighth throw. But the chances of getting less than that are better than 99 to one. You would have a 99.23 per cent chance of losing out. Unless the money you invest is meaningless to you, it is unlikely that you would ever sensibly take that kind of gamble.

The reverse applies to something like entering a major lottery. The risk of losing is very high – but the stake at £1 or £2 is a relatively small amount to most players, so they are prepared to take that high risk for a small chance of a huge reward, even though in simple economic terms it doesn't make sense to play. What the player is buying is not just the expected value but also the thrill of a possibility, the anticipation of what it would be like if a potential win occurred. Lottery players are, in effect, paying in part for entertainment, not just for the expected value.

Bernoulli's theory (which was itself much too simple to model real people) was that the utility was directly proportional to the amount of money and the value of the property you owned. To take a simple example – say your total assets were £10,000. Should you invest £5,000 of that with a 50:50 chance of doubling your money or losing it? The simple expected value gives no guidance. But the utility is a different matter. The point is that if you lose £5,000 you have lost half your assets. But if you gain £5,000 you have only added a third of your new net worth. At any particular level, a loss has more impact than a gain of the same size – but the more money you have, the smaller that impact becomes and the more likely you are to take the risk.

The law of large numbers

It was Daniel Bernoulli's uncle Jacob (they were quite a family) who came up with an idea that gives us the appearance of taming randomness – something that is often called the law of large numbers. It is beguiling because it works remarkably well with classical randomness, but is dangerous in the real world because it is no use at all with chaotic randomness. What the law of large numbers tells us is that as we get increasingly large numbers of a repeated classically random event, then the value observed will get closer and closer to the expected value.

Take the simplest example of tossing a coin. Assuming it's fair, we know that we should get heads 50 per cent of the time and tails 50 per cent of the time. This tells us nothing about the outcome of a single coin toss, but as we accumulate larger and larger numbers of tosses, we are more and more likely to come close to 50:50.

As a simple experiment, I tossed a coin ten times. The results were:

H T H H H T H H T T

Notice that out of the first five tosses, four of my results were heads. Nowhere near a 50:50 outcome. But over time, the result homed in on the expected value. Ten tosses are not enough to make it very likely that I would get 50:50 – it still isn't a large number. But the outcome, 60:40 is heading in that direction. One interesting observation in doing this exercise is that if I had decided to do an odd number of tosses I could never exactly hit 50:50 heads and tails – one of them has to be ahead, so some care needs to be taken with the size of the sample used.

Bernoulli used a slightly more sophisticated example of prediction. He imagined a large jar full of black and white stones, a jar that happened to have 3,000 white and 2,000 black stones in total. But we don't know this. We start to pull stones out of the jar and count up the different totals. In Bernoulli's example, the stone is returned to the jar after making the selection and the jar shaken up until the stones are randomly distributed. He worked out that after 25,500 stones had been checked, you could say with 99.9 per cent accuracy that the ratio of white to black was three to two, give or take 2 per cent.

The example Bernoulli gives may not be very practical, but it is the model on which many of our probabilistic predictions of the future and of chance are based. We assume that large numbers give us a safe ability to pin down randomness. The problem is, though, that in the real world chaotic randomness always has the potential of giving things a twist. For some reason, all the white pebbles could suddenly turn black, thanks to some unexpected chemical or nuclear reaction. Or, more plausibly, when we give the jar a vigorous shake to mix up the pebbles, they could smash through the bottom, leaving no pebbles in the jar at all. Chaotic randomness might not be numerically predictable, but we *can* predict that it will sometimes happen, whether it is for bad (as in a banking crisis) or for good (as in Harry Potter).

Distributing the outcomes

I've referred several times to the importance of distributions in classical randomness without really explaining what I meant. In classical randomness we can't predict what an individual event or outcome will be. So, for instance, drawing a pebble from Jacob's jar there is no way for us to say beforehand if it will be black or white. But we can say that the different possibilities are distributed in a particular way that will remain true. We could show the stones as a simple bar chart with two bars – one (white) of height 3,000 and the other (black) of height 2,000. Similarly the tossed coin has a distribution that is two equally sized bars, one for heads and one for tails. But that doesn't really tell us a lot about what will happen when we draw out a few stones, or toss a few coins. But we can get more information if we draw out a few stones repeatedly and plot the ratio of white to black.

Many examples of classical randomness obey a normal distribution, also known as a bell curve or a Gaussian distribution (though the great German mathematician Carl Friedrich Gauss did not discover the significance of the distribution, merely picking up on the work of French mathematician Abraham de Moivre). A plot of the values looks like a slice through the middle of a bell. In the centre of the curve is a peak. Either side it drops away, first slowly, then more quickly, until you are left with very low probabilities heading off to each side. When a normal distribution applies – which isn't the case with all classical random occurrence – then we expect to find most of the events or measurements to be in that middle section with increasingly few either much smaller or much larger.

The bell curve doesn't apply to Jacob's jar or coin tosses directly. But if I drew a fixed number of stones – 50 say – a repeated number of times, then the ratio of white to black in those sets of stones would follow a normal distribution, clustered around the average of $\frac{3}{2}$. If we want to know how likely a particular hypothesis is, then a measure of the spread of that curve called the standard deviation gives us a prediction (always providing we are dealing with classical randomness). So, for instance, when in 2012 CERN announced the discovery of a Higgs

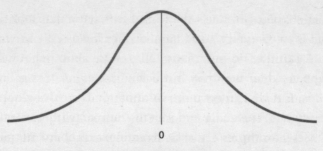

0

A normal distribution

boson-like particle with '5 sigma', what was meant by this is that for the results to have happened randomly without a Higgs boson, they would be five standard deviations from the mean (average). This equates to saying it would happen three times out 10 million attempts.

A simple example of a normal distribution would be the weights of mobile phones. Most would be clustered around the 100–120 grams mark. The average will be the centre of the curve, with tails heading off in both directions. A few will be ultra-light or seriously clunky, but the two 'tails' of the curve get to low probabilities very quickly. The chances of finding a phone weighing 10 grams or 200 grams is relatively low. There are lots of occasions in physics and nature where this kind of distribution does the job. It will help us, for instance, understand how we can predict the amount of pressure that comes from many billions of air molecules. But we always have to be sure that such a distribution really does apply, and that we aren't dealing with chaotic randomness.

If such a distribution does apply, then we can also expect regression to the mean. This was a phenomenon first observed by the Victorian scientist Francis Galton, and in a way it's just common sense. A simple way of looking at regression to the mean says that if we observe one item in a distribution and it has an extreme value, then observe another item, the second result will tend to be closer to the mean (the average value) than the first. So, for example, extremely tall parents will tend to have children who are closer to average height than their parents are. This isn't particularly surprising – if a particular value is extreme, which is a rarity in a normal distribution, it's most likely the next value will be closer to the centre of the distribution. This can be used to give some feeling of what will happen next when observing values over time that fit a normal distribution.

Regression to the mean is potentially dangerous, though, because it can make apparent causes appear out of nowhere. For example, in the UK it's not uncommon to put up signs at 'accident black spots' – stretches of road where there are frequent motoring accidents. The signs will say something like '27 accidents here in the last five years'. It is generally assumed that these signs reduce accidents because after the signs have been put up, the number of accidents at the location falls. But let's just see what would happen if the signs did nothing at all – in fact if no sign was put up.

Assume the accidents are actually randomly distributed around the road system. We wouldn't expect them to be nice and evenly spread out – in fact they wouldn't be random if they were nice and even, just like my dropped box of ball bearings. So in some place there will be clusters of accidents and in other places relatively few. There is no cause for these clusters other than the nature of randomness. Regression to the mean tells us that over the next time period, the chances are that a stretch of road with a cluster will come up with a number of accidents that is closer to the average. There will be fewer accidents than before. So by doing nothing whatsoever you will reduce the number of accidents. You *could* put up a sign, but you would be wasting your money. Of course there may be stretches of road where there is a cause for higher than average rates of accident. But we can't assume that there is such a cause just because there were a lot of accidents in one period.

In the last couple of chapters we have come across the foundations for our exploration of the improbable world. For classical randomness, probability tells how likely something is to happen and statistics let us combine many randomly behaving things to produce a predictable whole. But all the while the monster of chaos is lurking. Classical randomness only applies if the distinct random events brought together are independent of each other – like individual coin tosses. In the real world very few events are truly independent. This is one of the reasons the weather is so difficult to forecast – everything influences everything else, and the result is chaos. One surprising implication of our increasingly connected, social media-linked world is that it's harder and harder to find truly independent events and actions. Facebook and Twitter enable chaos to have a field day.

Before we see this real, chaotic random world in action, though, we have to take a step back into the Renaissance world, where the newly liberated scientists thought that they were in the business of removing randomness entirely to reveal a universe of clockwork regularity.

CHAPTER 5

The clockwork universe

As long as civilisations have existed we have attempted to perform a kind of science, imposing our mental patterns on the natural world. Initially these were primarily religious patterns. So, for instance, the rising and setting of the sun or the sudden impact of a bolt of lightning would be fitted into an explanatory scheme of gods riding chariots across the sky or getting irritated with humans and stirring things up with a good thunderbolt. But the Ancient Greeks, beginning with Thales who lived roughly between 624 and 546BC, gave us another model for thinking about the world.

When the later Greek philosopher Aristotle looked back at Thales he considered him to be 'the first founder of this kind of philosophy', where 'this kind' was a philosophy that tried to find a 'material cause' for things. Aristotle called Thales and his successors 'physikoi' or physicists to distinguish them from the alternative of 'theologoi' or theologians. What the early Greek philosophers did was not truly science in a modern sense, as they didn't care much about experiment and observation, but at least they were looking for patterns in the physical world that didn't depend on the intervention of the gods.

A good example would be the two Ancient Greek theories of the constituents of matter. Before the Greeks, stuff was just, well, stuff. Water was watery stuff, flesh was fleshy stuff and so on, pretty well indefinitely. There was no pattern for explaining stuff. But the Greeks came up with two potential patterns. One, mainly the work of the fifth century BC philosopher Empedocles, was that there were certain key substances – elements – from which everything else was constructed: earth, air, fire and water. The earlier physikoi had thought everything was based on a single fundamental substance, 'arche' (as in an archetype), but Empedocles was building his idea more on practical experience. When, for example, a piece of wood was burned, then fire and hot air (and sometimes oozy, sappy liquid) was given off, leaving earth-like ashes.

A different kind of observation produced the second theory. Two contemporaries of Empedocles, Leucippus and particularly his pupil Democritus, thought about cutting something up into smaller and smaller pieces. Eventually you could cut no more. What was left was uncuttable, a-tomos or atoms. These, they thought, were bouncing around in the void until they combined to take on the different forms of things that were found in the natural world.

Strangely, of the two, the one that was wrong was probably more like a true scientific theory. Admittedly both Empedocles and Democritus came up with a hypothesis, though in the Ancient Greek style, argument rather than experiment was used to decide between them. But the atomic theory didn't really tell you as much *about* stuff. In the original version, all atoms were made of the same substance, the arche, but had different shapes that meant they could only interlock with similar shapes. So depending on their shapes, there would be cheese atoms and water atoms and people atoms – this wasn't a pattern that gave much simplification to the various kinds of *stuff*.

By comparison, the theory of four elements, far from the truth though it may be, did simplify the picture, producing a clear pattern of four components that were mixed and matched to produce any substance. It might not have been true – in effect it got things back to front – but the four elements theory was a good example of applying a pattern to simplify and understand reality.

Come the Renaissance, the kind of thinking that the Greeks had pioneered was taking on a whole new dimension. The classical approach was still taught in universities, ossified from science to a belief system. But a new breed of thinker, typified by Galileo, was prepared to question the wisdom of the ancient authorities and to make new hypotheses based on observation and experiment. Galileo didn't just theorise from an armchair, he actually *did* experiments. Admittedly he probably never dropped his famous balls off the leaning tower of Pisa – this story seems to have been concocted by an assistant in Galileo's old age – but he made plenty of observations. And soon after him came the epitome of the new scientist (not called this though, as the name wasn't coined until the 1830s), that remarkable thinker, Isaac Newton.

The universe according to Newton

Newton's physics did more than provide us with laws of motion that were easy to use to predict what would happen, or a better understanding of how white light was made up of the colours of the rainbow. It provided a whole new level of pattern to bring the universe into a more understandable, more controlled system. By developing the mathematics to explain everything from the orbits of the planets to the way the tiniest speck of dust moved, Newton had done away with the need for heavenly powers to be present in the workings of the universe. Instead of worlds that were pushed around their orbits by angels and kept in place by the will of God, Newton gave us a kind of clockwork universe, where everything proceeded mechanically, predictably and with nothing truly random ever happening.

It's ironic, given the way that Newton's mathematics moved us away from the need for supernatural intervention in the day-to-day working of the universe, that at least one of his theories was itself attacked for being occult. In a superb burst of mathematical genius, Newton had produced his *Philosophiae Naturalis Principia Mathematica* – the chunky tome that includes his laws of motion and of gravitation. This book is often near unreadable to the modern eye, making heavy use of geometry where we would use far simpler algebra, but it certainly delivers the goods in predicting the way that objects move. One thing Newton failed to do, though, was explain *how* gravity did its stuff at a distance. The problem his opponents seized on was that Newton used the term 'attraction' to refer to the force that, for example, keeps the Earth orbiting around the Sun.

This seems perfectly normal now – the attractive force of gravity is part of our everyday thinking. But at the time, that word 'attraction' didn't have the scientific meaning we assume. It just meant the kind of attraction you get between man and woman, an animal reaction rather than a physical force. It seemed to his critics that Newton thought the Earth and the Sun (and for that matter, the Moon and the Earth) fancied each other. Just take the response to Newton's work of two of his great rivals, the Dutch scientist Christiaan Huygens and the German mathematician Gottfried Wilhelm Leibniz.

Huygens, who was in the habit of criticising Newton's ideas as a matter of course, something that did not endear him to the great man, was baffled by this use of 'attraction'. He threw up his hands at the 'theories [Newton] builds upon his Principle of Attraction, which to me seems to be absurd.' As for Leibniz, a superb mathematician with whom Newton had a massive plagiarism battle over who first devised calculus, the concept was retrograde. Leibniz dismissed the possibilities that two heavenly bodies could be attracted towards each other, calling it a 'return to occult quantities and, even worse, to inexplicable ones.'

They did not think the theory was occult in the modern sense of using black magic, but rather that it was something that was hidden with no obvious cause. As Leibniz was well aware, for an event to be caused remotely, something has to pass from A to B. To hear someone speaking across a room, a compression wave has to pass through the air from the speaker's vocal chords to stimulate the listener's eardrum. To knock a tin can off a fence, something has to cross the intervening space and dislodge it. You can't just look at it and make it happen. Yet Newton was saying that somehow planets and stars could influence each other remotely without anything passing between them.

Even Newton himself was not comfortable with this. Despite claiming in *Principia* that he 'framed no hypothesis' for how gravity worked, he actually had ideas that there was some sort of invisible stream of particles constantly flowing between the massive bodies to make gravitation happen. Until Einstein came along, the best-supported theory for the cause of gravitation would be something mechanical of this kind. The idea was typically that there were streams of special gravity particles flowing throughout the universe. As a heavy body would reduce the number of particles reaching a nearby body by screening it, the result would be that the body was hit by fewer particles and would feel pressure to move towards the heavy body. It was attracted. There were details to work out – why gravitation depended on mass, not size, for instance – but at least it was a theory that did away with that damnable occult force.

In the end, though, Newton's numbers worked. They accurately predicted the behaviour of moving objects and orbiting bodies. They described a mechanical picture of the universe that was hugely successful in its match to reality. Clearly there was a cause, a reason for the attraction to take place, and at some point that would be understood, but in the meanwhile it would be churlish to ignore Newton's mathematics, which gave such an elegant and accurate prediction of what really happened. For many pragmatists it was a case of 'Don't worry about *why* it works – it does, so let's use it.'

Probably the greatest of such pragmatists was Pierre-Simon, the Marquis de Laplace. This French mathematician and scientist was unusual at the time in his enthusiasm for Newton's work. Both because of Newton's arguments with Leibniz over calculus and his idea that light was made up of particles rather than the waves suggested by Huygens and Descartes, the Englishman was treated with some caution by many philosophers from mainland Europe. They accepted the brilliance of his work on motion, but were a little more wary about plunging wholehearted into the detail of his science. Laplace, though, a brilliant mathematician in his own right, took Newton's inspiring insight into the workings of the universe to a whole new plane.

No need for that hypothesis

Although Newton was aware that his explanation of motion and gravity should be sufficient for the universe to function without intervention, he was also a very religious man and expected a role for God in keeping the mechanics of reality on track. He found one in the stability of the universe. The problem was that by simply applying Newton's maths, you might expect the universe to collapse. Newton first imagined what the universe would be like if it were finite. If you looked at a planet or star near the edge of that finite universe, it would feel a strong attraction towards the centre of the universe, because there were many more bodies in that direction. So the universe should collapse from the edge inwards.

Newton got around this apparently inevitable breakdown of everything by assuming that the universe was infinite (a controversial concept in his day). That way it would have no edge, and wherever a heavenly body was located within the universe it would have plenty of other bodies attracting it in all directions. However there was still a problem. Such a universe could only be stable if everything was in exactly the right position, as originally positioned by God in the creation. Inevitably, some bodies would get just a little out of position – at which point they would start to drift, and the collapse would come again. It would just take a lot longer this way.

The only way Newton could see to get around this was to have God ever-present, gently prodding things all the time, ensuring that everything stayed in the right place and that collapse would not occur. For Laplace, though, God was an unnecessary universal component, because there was no room for drift. The mechanical perfection of the universe should extend to each and every body and motion. According to legend, when Napoleon asked Laplace why there was no mention of God in his philosophy, Laplace is supposed to have replied 'I had no need of that hypothesis.'

Laplace believed that there was no such thing as a random influence and so there should be no need for constant corrections from a caretaker God. He truly envisaged the universe as one vast clockwork mechanism. If someone could understand every single universal cog and link, if that person had perfect data on every particle in existence, then they should be able to foresee the behaviour of everything in existence into eternity. There was no room in Laplace's universe for uncertainty.

Laplace was very clear that with sufficient data, unlimited understanding and the ability to analyse a vast quantity of information, it was possible to predict the way the entire universe would run. He wrote, 'Given for one instant an intelligence which could comprehend all the forces by which nature is animated and the respective situation of the beings who compose it – an intelligence sufficiently vast to submit these data to analysis – it would embrace in the same formula the movements of the greatest bodies of the universe and those of the lightest atom; for it, nothing would be uncertain and the future, as the past, would be present to its eyes.'

At first glance there is something inevitable and grand about Laplace's vision. We know that everything, even the human body and brain, is made of individual, inanimate atoms. At the level of these atoms, there is no such thing as 'life' or 'consciousness' or 'decisions'. In Laplace's world, these atoms obey physical laws that determine how they will change from moment to moment. Capture every bit of data for every single atom in existence and you have written down the computer program that runs the universe.

With enough resources, you should then be able to run that program forward indefinitely and you would then predict exactly what would happen for all time to come. It is indeed a grand vision – but also a depressing one. In Laplace's universe there is no room for free will. What will happen now or in 100 years' time was set at the beginning of the universe when everything was given its initial conditions. From then on each atom trudges its way through its inevitable path, que sera sera, with no hope of intervention or change, no random influence or disturbance from the grand design. Forever.

The only apparent escape clause from an existence where human beings had no true choice about what they did was if you believed a concept that dated back to the Ancient Greeks. It was given the form that Laplace would have recognised by 17th-century French philosopher René Descartes. The idea, known as dualism, sees a human being composed of two independent components – a body that is mechanical and material, and a mind (which for religious purposes could equally be regarded as a soul) that is supernatural and immaterial. This mind part is 'supernatural' in the sense of being outside of nature, with no horror film implications.

If there were such a separation between mind and body, then it's possible that the physical aspects of the universe would follow Newton's mechanical necessity, but the thinking part of human minds, existing separately and outside of nature, would not be straitjacketed and could steer the physical world, breaking the bonds of the clockwork universe. To take control, the mind and body would have to be somehow tied together. This is the weakest point of the theory because it requires an interaction between the natural and the supernatural, the physical and the immaterial. Descartes suspected the pineal gland in the centre of the brain provided the linking point between the two parts of a human being, something we can now be sure isn't the case.

Most people throughout history – including the majority of people alive today – have held a dualist view, because it is the natural, commonsense one. (This doesn't make it right. The natural, commonsense view also says the Sun travels around the Earth.) Even the most sophisticated modern philosopher can't avoid thinking of his or her 'self' as something separate from the body that it controls. If you take a moment to think about it, the same is true for you. All of us inevitably imagine a sort of mental being, probably located between our eyes, that is somehow pulling the levers to make the physical body work, via the intermediary of the brain. However, the majority of present-day scientists believe that there is no such duality, and that the mind is simply a function of the chemical and electrical functions of the brain.

It ought to be stressed that this idea of human beings as 'meat machines' cannot be proved scientifically, and as yet we have no good answer to the nature of consciousness. It is also true that to dismiss duality runs contrary to pretty well all the world religions – certainly any that consider us to have a soul or to be capable of an existence after death. However, if there is no duality – and supporters of the mind-as-function-of-brain view would point out that there is equally no evidence for a separate mind, other than our subjective feelings – then with Laplace's world view, there is no escape from the iron grip of the predictions of physical laws. (The approach proffered by Laplace is often described as 'deterministic' as the initial conditions and the rules of the system determine entirely how things will play out.)

Without doubt, the precision of Newtonian mathematics seemed to provide a new level of insight into the workings of the universe. Of course, we humans were not Laplace's 'intelligence which could comprehend all the forces by which nature is animated and the respective situation of the beings who compose it.' We have neither the data nor the computational skill to predict the moment-by-moment progress of the whole universe. But Newton seemed to promise that in principle this was possible. For the moment, though, it was best to start simple.

If you look at a very simple gravitational system with just two bodies interacting – the Earth orbiting the Sun, or the Moon orbiting the Earth, say – and take away all other influences, Newton's maths does indeed (within the limitations that would later be revealed by Einstein in general relativity) give us the means to predict *exactly* what will happen. For all time. And no doubt the enthusiasts for determinism thought that it was just a matter of adding in body after body, detail after detail, until we had our master plan for the whole universe assembled. But this magnificent progress of the imagination through the workings of a clockwork universe was brought up short by reality.

Just add one more object into your two-body system and everything goes horribly wrong.

CHAPTER 6

Just three bodies

The natural starting point when trying to make calculations of the movements of a 'clockwork' universe is to consider the Earth and the Sun. After that? Why not add in the next most obvious body in the sky, the Moon. So now we have three bodies. The Earth orbiting the Sun, the Moon orbiting the Earth. A simplification of the real solar system, it's true, but clearly more complex than predicting the behaviour of two bodies. More number crunching would be required. And yet it was hardly a major challenge. At least, so it seemed until scientists and mathematicians tried to make it work.

Newton was the first to give the problem serious consideration. He made some initial progress, but was aware that it wasn't such a trivial problem as it might first seem. Clearly both the Earth and the Sun influence the movement of the Moon. However, to simply calculate that impact assumes that the bigger bodies are not themselves influenced by the Moon. But they are. It has a gravitational pull on them. This will displace the Earth and Sun from where we thought they were in terms potentially of both position and the way that they are moving. But that will mean a different impact on the Moon – and so on.

When you only have two bodies to deal with, it's easy enough to account for their effects on each other, but with the three, the perturbations each makes on the orbits of the others leads to a sort of chaotic randomness that is

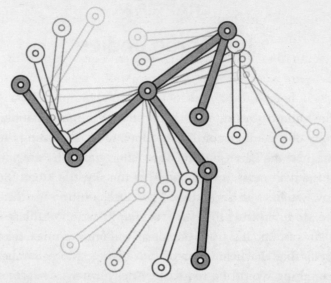

A chaotic pendulum, with multiple jointed segments that influence each other, provides a similar demonstration to the chaotic interaction of three bodies.

difficult to predict exactly. In his *Principia*, Newton split off and worked on elements of the problem, but didn't take it on in its entirety. So, for instance, he made a stab at calculating 'the forces of the sun that perturb the motions of the moon', picking out that particular influence, without bringing the whole together.

In practice it was discovered pretty early on that dealing with the three-body problem resulted not just in an outcome that was difficult to predict, but an impossibility. Because there are soon many different states that could originate from the same starting point, it isn't possible to exactly predict what is going to happen. Instead, mathematicians dealing with the three-body problem (and even more with our real solar system featuring significantly more than three bodies) have to resort to approximation.

With enough computing power, these approximations are very good. When the Apollo missions were dispatched to the Moon they were making use of basic Newtonian physics with good enough approximations to handle the flight of the Apollo vehicle, including the influences of Sun, Earth, Moon and outer planets, particularly the massive Jupiter. This was fine to deliver the lunar landers to the required part of the Moon's surface and to get them back to Earth. But it wasn't Laplace's perfect prediction of the behaviour of a mechanical universe. Extrapolate it further and further into the future and the predictions given by the maths would get further and further from reality. It was only workable as a relatively short-term approximation.

Things got even worse when relativity came on the scene. Relativity itself dates back to Galileo and simply states that for movement to exist we have to know what our 'frame of reference' is – with respect to what, exactly, are we moving? It might seem obvious what is moving and what isn't, but that's only because we typically assume the frame of reference of the Earth. As you sit in a chair or wherever you are reading this, you probably aren't moving with respect to that seat. But the seat could be moving with respect to the Earth if you are on a train or a plane, for example. And for that matter, the Earth itself is moving with respect to the Sun or other points in the universe. So, Galileo told us, the whole concept of movement has to be *relative* to something. That's relativity.

Relativity becomes special

Einstein took this one step further in his development of the concept of special relativity in 1905. Thanks to the work of Scottish physicist James Clerk Maxwell it had been discovered that light was an interplay of electricity and magnetism. Moving electricity created magnetism; moving magnetism created electricity. But this could only haul itself along by its own bootstraps if it moved at one particular speed, the speed of light – in a vacuum around 300,000 kilometres per second.

This presented Einstein with an interesting challenge. He imagined floating alongside a sunbeam. As far as the imaginary floating Einstein was concerned, relativity said the sunbeam wasn't moving. And that meant it didn't exist – because it wasn't going at the right speed for electricity to create magnetism to create electricity and so on. In fact whenever anyone moved, all the light around them should disappear – which seemed bizarre.

This paradox forced Einstein to make a leap of imagination. What if, he thought, light was unlike everything else in the universe? What if light always moved at the same speed, however you moved towards it or away from it? Light, effectively, would ignore relativity and just do its own thing. That way, however you move, whatever your frame of reference, light would cruise on regardless.

This move had the desired effect of making it possible for light to exist in our relativistic universe. But there was a cost – a big one. If you plug the behaviour of light into the basic equations of motion and energy derived from Newton's work, it transforms the moving object. Whenever anything moves, three things happen to it. Firstly, the moving object increases in mass. This happens at any speed, but it is hardly noticeable until you are travelling a good-sized fraction of the speed of light. However, once you do get to very high speeds the mass absolutely shoots up, so that as you approach light speed, mass tends to infinity.

The way that mass increases is one of the reasons it is impossible to simply fly faster and faster until you exceed the speed of light. As you get close to light speed it takes more and more energy to make you accelerate, because the energy required is proportional to your mass. It would require an infinite amount of energy to work through the light speed barrier (though there may be ways to get around it).

As well as an increase in mass we also see a moving object getting shorter/thinner in the direction it is moving, squashing up more and more until it becomes infinitely thin at light speed. And, most strangely of all, the time on the moving object gets slower and slower until it stops at light speed. These are all relativistic effects. This is what happens on the moving object as seen from the place with respect to which it is moving. If you *were* that object you wouldn't notice any of these effects happening to yourself. In fact, because all movement is relative, you don't see yourself moving at all, it's the world that moves around you, and it's everything else that you will see experiencing these effects.

The eternal triangle
So let's go back to our three-body problem. The bodies in question are moving – that's the whole point of the

problem. So because of that movement, each will see the others as having increased mass. This effect is pretty well negligible at the relative speeds of the Sun, Moon and Earth – but it is there as far as Laplace's perfect observer is concerned, and it means there is an added shift in the already impossibly slippery calculation. It's even worse for bodies that are moving at near light speed, where the impact of movement on mass becomes really significant and throws the whole set of equations used to describe what's going on into extra depths of complexity. Even more so with special relativity in play, the only feasible approach is one of approximation, though it is possible to make the approximated values more and more accurate.

It might seem strange that it is possible to make an approximation to pretty well any level of accuracy but not to be able to take the final step and come up with the ultimate, exact value. To see what is happening it is worth contrasting two very different infinite numerical series – the result of combining a list of numerical values that has an infinite set of entries.

A very simple infinite series is the sum $1 + \frac{1}{2} + \frac{1}{4} + \frac{1}{8} + \frac{1}{16} + \frac{1}{32} \ldots$

We can say that the sum of this series, the result of adding all the values together, tends to a value of 2. If we had the whole infinite set of fractions (which in practice we can never have in the real world), we would get an exact value of 2. So even though we can't do the actual sum, we can come up with an exact prediction of where the series is heading. We can provide a perfect forecast for the outcome of the sum even though we can't complete it in reality. Consider another relatively simple infinite equation, though:

$$\tfrac{2}{1} \times \tfrac{2}{3} \times \tfrac{4}{3} \times \tfrac{4}{5} \times \tfrac{6}{5} \ldots$$

The sequence is very easy to predict, for as far as you like. At one step the bottom part of the fraction increases by two and on the next the top part increases by two. And in a sense I can tell you exactly what the outcome will be, just not as a number. The result of including the whole infinite set of entries is $\pi/2$ (where π, or pi, is the familiar ratio of a circle's circumference to its diameter). The more values you include in the sequence, the closer you come to getting an exact value for pi (or more accurately half of pi). But we can't say what numerical value the sequence is heading for. It isn't ever exactly predictable. We can approximate it to any level of accuracy – and pi has been calculated to millions of decimal places – but with this kind of calculation we can never forecast the eventual destination of the finished sequence. It's what's called a transcendental number.

The same kind of inability to get to the final value goes for the three-body problem (or the 'n body problem', as it is known when there are more than three objects involved). With enough computing power, we can calculate the values to pretty well any desired degree of accuracy, but we can never forecast what the exact final results will be.

As we will see later (see page 147), relativity wasn't the only bit of modern physics that would throw a spanner in the works of Laplace's predictable universe. Quantum theory added its own random baggage to the ability to make predictions. Even at the basic level, the three-body problem becomes more complicated when dealing with the interaction of three quantum particles. The problem here is that it is perfectly possible for new particles in pairs – matter and antimatter – to briefly pop into existence, then soon after collide and annihilate, returning to pure energy. This is all very well, but it means that at certain points in time there are more than three particles present, subtly shifting the positions of the original three. There are added perturbations that make any attempt to predict behaviour ever more an approximation.

There were some partial solutions to the traditional three-body problem. The French mathematician Joseph Louis Lagrange worked on a problem where two of the bodies were much larger than the third. In such circumstances, it is possible to identify points where the gravitational pull of the two large bodies is exactly the same as the amount of force needed to keep a third object in orbit. For these specific points, known as the Lagrangian points, there is an exact solution to the equations. There are five such points in total – two in positions forming an equilateral triangle with the heavy bodies, one between the two bodies and one either side of the two bodies on the extension of the straight line between them.

The Lagrangian points in the Earth/Sun system are of particular interest. In a near-circular orbit like the Earth's around the Sun these are pretty well true points (they spread out to be fuzzy patches in a more elliptical orbit) and they provide a stable place to park a satellite, orbiting around the Lagrange point. In practice, the Earth/Sun

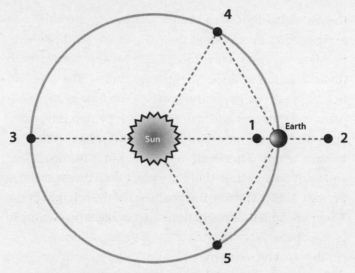

The Lagrangian points for the Earth/Sun system

system is more complex because there are influences from the other planets, but to a good approximation the Lagrangian points, particularly the 'L4' and 'L5' points on the extremes of the two equilateral triangles, are islands of stability in a chaotic gravitational environment.

In the general three-body problem, once we add in an extra body to the initial two we arrive at a state where subtle differences in starting conditions can result in major changes further down the line. It becomes harder and harder to make predictions over time.

Exactly the right conditions for an attack of chaos.

CHAPTER 7

Chaos!

Imagine standing above a mountain range on a specially constructed platform, with a huge boulder clasped in the grapple of a crane. Ahead of you is a sharply crested ridge. Either side of the ridge, as you look down, is a village. The drop from here is precipitous. If your boulder fell on either village it would cause death and destruction. Release the boulder with its centre of gravity one centimetre to the left of the ridge, towards the northern village and the destruction there will be terrible. The other village will be untouched. But just move the boulder two centimetres to the right and it is the southern village that will be devastated. Such a vast difference in outcome from such a tiny change in starting point. A move of just two centimetres deciding who will live and who will die.

It is (thankfully) an unlikely image in practice, but it illustrates well the essential nature of chaos. A very small variation in initial conditions – in this case the exact position of the boulder over the ridge – a change that it would be difficult indeed to see if simply lining up the boulder over the ridge by eye, makes a huge difference to the outcome. Predicting what is going to happen with any certainty becomes impossible.

This concept of chaos emerged from early computer work on weather forecasting. The American meteorologist Edward Lorenz had produced a simple model (simple by modern standards, at least) of the way that weather patterns develop over time. As is often the case with computers, the printout of the results worked at slightly less accuracy than the numbers used in the actual model. So, for instance, if a value that was being worked with was 0.349456, the computer might, to keep paper usage down, print out 0.349.

Lorenz wanted to recreate a run of his model from part way through, so he carefully input the values he had on paper as starting positions for the model. By using those rounded values on the printout he was entering something ever so slightly different from the figures the computer had previously been working with. But Lorenz assumed the impact would be trivial. It's as if you were setting up your sat nav, and instead of locating the exact position of the car in the drive, you put in the coordinates of the street outside the drive. A trivial difference that should not influence the outcome. Yet as Lorenz's model ran forward in time, it rapidly diverged from the original forecast in a massive amount of detail.

A very small change in the initial values – in this case, the difference of a few extra decimal places, at most varying by one thousandth – made a huge difference in the outcome. Lorenz would never have seen that coming. The kind of information he was typing in – air pressure or temperature, for example, would certainly not have been measured to anywhere near this level of accuracy, so it seemed irrelevant that he should use truncated values from those the computer had been working with. In reality it made all the difference.

Lorenz was personally responsible for our most dramatic image to portray the impact on outcome of a small change in starting point when he published an academic paper entitled *Does the flap of a butterfly's wings in Brazil set off a tornado in Texas?* This so-called 'butterfly effect' captured the public's imagination. As it happens, the real answer to Lorenz's question was 'No.' The effect of a butterfly's wings is so small that it would tend to be damped out by the system rather than amplified – and tornadoes are typically relatively localised weather systems – but the basic idea was valid, and was one that would have a huge impact on the science of weather forecasting, arguably our best attempt to tame chaotic randomness.

Oddly, the initial response to the discovery that the weather developed in such a knife-edge fashion was one of elation. Weather forecasting is really only a second-best activity. Ideally what we want to be able to do is to control the weather and have it behave the way we want, not simply to forecast it. The eminent mathematician and computer scientist John von Neumann was delighted at the discovery of the delicately balanced nature of the weather because he initially had in mind the sort of behaviour described at the start of this chapter with a boulder hanging over a sharp ridge. He imagined that if the weather was so finely balanced, it should be easy to give a weather system a little nudge one way or the other to get the outcome he wanted.

What von Neumann didn't grasp initially was that the weather was not so much a single mountain ridge as a whole matrix of mountain ridges. Yes, it would be possible to influence a first finely balanced outcome – but that would trigger a whole new set of finely balanced outcomes, each with its own narrow ridges. This wasn't like pushing the boulder to one side or the other, it was more like giving a ball a slight sideways nudge in an immensely complex pinball machine.

The unpredictables

The broad mathematical theory that would describe this kind of behaviour was given the label 'chaos theory'. Many large systems that have a number of factors influencing outcomes have this kind of chaotic behaviour, whether we're thinking of the weather, the sales of books or the behaviour of the stock market. In principle this is the sort of randomness that could be tamed by Laplace's all-seeing, all-knowing entity. It is not truly random, because if you really did have total knowledge to the very last decimal place and enough processing power, you *could* predict exactly what was going to happen. But because in reality we always work with simplified models – often highly simplified – we don't see the dramatic outcomes of chaotic randomness coming, and as a result they can and do take us by surprise time and time again.

We should really have expected to discover this chaos in nature. It is what we experience as everyday life. Think of the old verse about losing a horseshoe nail that is sometimes linked to the death of Richard III at Bosworth Field:

For want of a nail the shoe was lost.
For want of a shoe the horse was lost.
For want of a horse the rider was lost.

For want of a rider the message was lost.
For want of a message the battle was lost.
For want of a battle the kingdom was lost.
And all for the want of a horseshoe nail.

Everyone is aware of the unfolding consequences of a small action on our future. From fairy tales, where a small kindness to a poor person who turns out to be a prince or a magician results in great rewards, to modern stories like the film *Sliding Doors* where two possible outcomes are followed depending on whether or not the central character played by Gwyneth Paltrow catches a tube train, we are familiar with the idea that even small decisions have consequences and we can rarely be aware just what will ensue. The consequential implications of our everyday actions are just as much a chaos effect as anything studied in weather, book sales or the stock market, but somehow we seem more comfortable with the fact that we can't predict our own futures.

The fact is, if we are uncomfortable with randomness, we are doubly uncomfortable with chaos. At least classical randomness is predictable across a large enough sample. Chaos runs away with itself, leaving predictability behind. The increasing complexity of the input of chaotic randomness to a weather forecast builds over time. Typically a forecast for one or two days can be surprisingly accurate. Up to five days out there is a fair chance of getting things right, although in practice the forecasts are often modified as time progresses. But get any further and things go downhill. In fact, bizarrely, at nine days or greater, forecasts are typically worse than simply basing your prediction on the kind of weather expected in a particular location at that time of year.

This seems to be quite an indictment of our meteorologists. The forecast at, say, ten days is not even as good as what you can do by understanding the basics of the local climate, it is actually *worse*. The problem is that the forecasting systems attempt to model the way that feedback in the weather causes dramatic change. Unfortunately these feedback loops don't just pick up on the accurate data that is fed into a forecast, they also amplify the errors, so that the models run away from reality in a drastic fashion after a sufficiently long period of time. It is particularly strange that some weather sources provide ten- to fifteen-day predictions despite being aware of this, though it's not clear if they do still use true forecasting models or simply fall back on an expectation for that location at that time of year.

If ten-day forecasts are frequently poor, it's not surprising that trying to forecast a month ahead leads to dire results. American operations research and economics expert Kenneth Arrow became aware of this when involved in weather forecasting for the US Air Force during the Second World War. A team of forecasters were attempting to predict the weather a month ahead, something we now know that chaos makes impossible.

Because Arrow had a mathematical background, he analysed the forecasts and found that they were no more accurate than picking a forecast at random and somewhat worse than guessing using the local climate as a guide. Based on his observations, the forecasting team attempted to get out of providing the useless forecasts.

The response was, 'The Commanding General is well aware that the forecasts are no good. However he needs them for planning purposes.'

Strangely, some commercial weather forecasts are intentionally less accurate than those provided by governmental meteorology departments. This is not because the commercial forecasters have worse data – they can usually access the same sources – but by shading the predictions in a particular way, the forecasts can be made to look more appealing. There seem to be two main factors in play. Firstly, there is a tendency to avoid a balanced outcome. If, say, there is a 50 per cent probability of rainfall, commercial forecasters will tend to shift it away from that 50:50 judgement, to avoid seeming indecisive.

The second commercial weather forecasting trick is a tendency to increase the chances of rainfall above the value predicted by the models. The reason for this deception is simple. It is worse to get your forecast wrong by saying it won't rain, leaving your customers with an unexpected soaking, than it is to get it wrong by saying it will rain and subsequently have them find they are enjoying unexpected sunshine. If you over-predict rain, you are less likely to get irritated customers, who blame you for the drenching. Who is going to complain if it's sunny unexpectedly? No one but a few rain-starved farmers. Psychology triumphs over meteorological accuracy, at least for forecasters who expect to be paid by the consumer, when it comes to the impact of this particular kind of random influence on our lives.

You might think that, with chaotic randomness at its heart, we might as well give up any attempt to make weather forecasts at all – but in reality, short-range fore-casts have improved vastly in quality over the last 20 to 30 years. At the time of writing it is 25 years since the UK was hit by one of the biggest storms on record, a hurricane that was not predicted by the Met Office. This would be very unlikely to happen now. This is because the forecasters have embraced the chaos, rather than pre-tending that it doesn't exist. Traditional forecasting never took chaos into account, making predictions on the same kind of statistical basis as we might use for forecasting the outcome of a roulette wheel. A modern weather fore-cast is quite different.

Aware of the variability resulting from small vari-ants in starting conditions, weather forecasters run an 'ensemble' of models, each with slightly different initial conditions. They then group the predictions together and use them to make those probabilistic forecasts. If, for example, there were 50 runs of the model with slightly different starting conditions, each as likely as the next, and 30 of them predicted rain, but 20 said that it would remain dry, you could say there was 60 per cent chance of rainfall. The actual translation of the ensemble of model runs into a specific forecast is rather more complex, but this gives a feel for what is happening.

The next bestseller

The weather is a relatively well-understood system. It has been studied in detail for many years and we have vast amounts of data collected on both the kinds of obser-vation that are made and the resulting weather. And by comparison to many human-influenced systems, it is relatively simple. If we look at another chaotic system, the sales of different books, we are faced with a much more difficult scenario to forecast. We have the figures on the outcomes – how many copies of each particular book sold at what time – but we have very little data on many of the variables that go into making the system work. It is hard even to specify what those variables might be.

Influences on book sales will clearly include the amount of marketing that is done, visibility of books in stores, how much prominence key online retailers like Amazon give, the media coverage, word of mouth observations, social media, available disposable income and more. But we are dealing with such a complex system, involving all the book buyers, bookstores and publishers as different variables, that making any attempt at forecasting is indistinguishable from guesswork.

As an example of the complexity of predicting what might happen to even a single book, imagine an ebook that is reduced from its usual price to 99p for a limited period, as part of a special promotion by an online retailer. It might seem that there should be a simple relationship between price and sales, but it doesn't work like that. With this sort of promotion, sales can increase far more than the shift in price would suggest. A big factor in this is improved visibility. Not only will the book be more obvious because it is in a special deals section, it may also rise up the ranking to be one of the bestselling ebooks, getting further exposure this way. The feedback loops in place make it hard to predict what is going to happen. Now add in all the complexity of dealing with not one book but all of them, not one retailer but all of them, and a much wider buying public than a particular ebook store. Once more, like the weather, small variants of the initial variables can result in huge differences in outcome, but we don't even know what those initial variables are in detail for book sales. And there are vastly more of them. An ensemble forecast would be inconceivable.

As the mathematics of chaos was developed it was realised that it applied to more and more situations, spanning a wide range of scientific disciplines. Take the population of a new species in a particular ecosystem. Once biologists got over their initial fear of mathematics, the assumption was that a population would typically experience initial rapid growth followed by a series of oscillations up and down as the population headed for a state of equilibrium. Admittedly, when they tried to plot the models of this kind of behaviour they didn't always reach that nice steady state, but the assumption was that they would get there eventually.

However, once serious computing power and decent maths were available, it was discovered that after initially going through a phase of relatively stable oscillations, rather than settle to a steady state, with a high enough level of fertility you would get a chaotic dive up and down of population as the ability to reproduce vied with ability to find enough to eat. Effectively what is happening is that the possibilities for change split into two, these themselves split into two other options each, the whole happening faster and faster until the end result is chaos.

A similar phenomenon in a very different area drew the literal poster boy of fractals and chaos, the late Benoit

Mandelbrot, into the fray. Mandelbrot was a young mathematician working at IBM's research centre in Yorktown Heights, New York. He had spotted an oddity in the way income distribution did not fit a conventional mathematical distribution, and was shocked to discover exactly the same kind of unruly behaviour in the distribution of cotton prices. These prices were ideal to study as they were well documented, going back many years.

What mathematicians and economists had assumed until then was that the behaviour of something like cotton price over time would be a combination of long-term responses to the big picture – the economy, new trends in use of fabrics and so on – and a short-term, random fluctuation up and down. These fluctuations were expected to fit a normal distribution – but they didn't. There were far too many extreme jumps. The traditional approach had been to ignore the 'noise' of the random fluctuations and concentrate on the trends. But Mandelbrot realised that this analysis got things back to front. In what would later be identified as chaotic data like this, it was often the dramatic spikes that dominated. A normal distribution simply misrepresented what was going on.

As Mandelbrot examined the data he made a discovery that would lead to most of his work on fractals – an area that is primarily out of the scope of this book – but which also explains what lies beneath chaotic randomness. What Mandelbrot found was that the changes in cotton price were independent of scale. Any particular change occurred randomly, without any way to predict it. But if he looked at the changes over a week, or a month, or years at a time, there was the same broad shape to the way changes went up and down. This didn't make it any easier to make a prediction for a specific change in price, but did make it possible to assess what should be seen in the broad distribution of the price changes over a period.

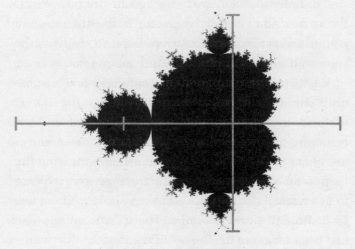

The Mandelbrot set

In the early days of the boom in understanding of chaos theory it was thought that this new mathematics, with its icon of the Mandelbrot set graphic (a plot of a chaotic system that exhibits a scale independence like cotton prices), would transform the way we understood the world around us. Chaos was seen in everything from a dripping tap to the movement of the stock market. Writing about the development of chaos theory in 1987, James Gleick commented, 'The most passionate advocates of the new science go so far as to say that twentieth century science will be remembered for just three things: relativity, quantum mechanics and chaos.'

While these enthusiasts were right about the span of chaos, they were wildly over-optimistic about the impact of chaos theory to do anything useful. The trouble is, the theory is great at telling us what we can't ever effectively predict. But it doesn't give us any greater understanding of, say, how the stock market will change tomorrow. It just gives us the insight that it's pointless to try and predict it. A useful insight indeed, but not one that is going to carry forward our understanding of the universe in the way relativity or quantum theory has.

When we are dealing with chaotic randomness it is very difficult to get an overview. Classical randomness is different. Although each random occurrence is totally unpredictable in its own right, at an overview level we can often make very good predictions about what's going to happen. This is where statistics has come into its own as it has got more and more sophisticated.

CHAPTER 8

Statistical substance

Statistics – ah, blessed statistics. You have probably come across the quotation 'There are three kinds of lies: lies, damned lies and statistics.' It is often attributed to British Prime Minister Benjamin Disraeli, but he was just quoting Mark Twain. (At least, he said that he was quoting Mark Twain, though no one can find evidence that Twain ever made the remark in the first place.) We are naturally suspicious of statistics. Yet used properly, they give us a firm handle on classical randomness.

When a scientist is trying to understand what is going on in a gas at the level of individual molecules, for instance, she faces the three-body problem writ large. There aren't just three bodies to deal with, but many billions, all interacting at some level. It simply wouldn't be possible to assess what is happening across the board by working out the path of each molecule. But instead, a statistical approach enables her to take an overview, to tame the randomness of the behaviour and pull everything together. Provided you have classical randomness to deal with, statistics is an immensely powerful tool to understand what is going on.

The reason Disraeli and the rest of us remain suspicious of statistics is twofold. As we have seen, the term comes from the same source as 'state'. Statistics originally derived from the first attempts at censuses and surveys of what was happening in a country. They had an inevitabe tie-in to politics, with all the distrust that engenders. There is an unavoidable feeling of having Big Brother looking over your shoulder whenever you become a statistic. The other issue is that statistics are so easy to misuse – as will become clear in this chapter – either intentionally or accidentally.

One common problem is knowing where a statistic has come from. Sometimes statistics, like those from a census, are produced by looking at the whole population, but frequently this is too difficult and too expensive a task to perform. The last UK census, taken in 2011, is thought to have cost around £500 million to compile – this isn't something that even a state can run to very often. So instead statistics are deduced from a sample, a subset of the population. Those compiling the numbers find out about a segment of the population they can get their hands on (traditionally this is a population of people, but it could just as easily be a population of molecules in a gas), then multiply up the results to provide the big picture. Such sampling is an imperfect science at best.

Even if it's done properly, sampling inevitably makes the result less certain than it sounds. You might hear on the news (a frequent purveyor of dodgy statistics) that 460,000 migrants entered the UK in a particular year. It would be reasonable to think that these numbers are the result of adding up all the individuals who turn up at passport control, but it isn't done that way. The numbers originate from something called the International Passenger Survey. A handful of individuals go out and interview the passengers at airports, ferries and the

Channel Tunnel. The figures are typical based on around 2,000 individuals a year.

When they say 460,000, they really mean it's probably between 433,000 and 487,000 – when you look at the detailed statistics you are told that in just 5 per cent of cases would the actual number be outside those bounds. But most people won't look at the actual figures; they will take the 460,000 as fact. All the statistical analysis can do is to provide us with the best guess, given the tiny fraction of the many thousands of travellers that are actually interviewed. This isn't a bad thing – but it can be dangerous if we don't know where that 460,000 number came from.

Sample selection

Whether by accident or intent, the way that a sample is selected can make all the difference as to how the statistics turn out. Let's say, for instance, you wanted to gather information on what make of car members of the public were likely to buy next time they went to a car showroom. If you deliberately wanted to bias a sample towards your own brand, you could just send the survey to people who had already bought your car in the past. They are more likely to go for your cars (unless your cars are terrible) than another manufacturer.

That is a deliberate attempt to mislead, but you could also bias the sample accidentally. For example, if you take a survey online, which many polling companies do, you immediately have a sample that is not representative of a population as a whole. In the UK in 2011, 77 per cent of households had internet access – but that left 23 per cent who didn't and who could never get access to your survey*. The chances are that internet access is related to factors like age and income. That means that those who are online will have different buying characteristics to the general population. So your survey will be biased. Similarly, if you only surveyed people working in the banking sector of a capital city you would get a very different picture of car preferences to people at a job centre in a provincial town.

Some such misuses could well be deliberate, but it is also possible to misinterpret statistics through carelessness. A good example was a British newspaper that announced a few years ago that the effect of raising the retirement age for men from 65 to 67 would be that one in five men who would previously have had a pension would now miss out. These men would die before they received their pension. This was, in fact, rubbish. One of the reasons that the stupidity of the statistic doesn't stand out is that the actual values are hidden behind that 'one in five'. Whenever statistics are presented as

---

* If you are paying attention you will realise that the statement '77 per cent of households had internet access' is also a survey based on a sample, so may have limited accuracy. Dive into the detail and you will discover that this headline figure is derived from the National Statistics Opinions Survey, a multi-purpose survey run by the UK Office for National Statistics. According to the background notes, 'The Opinions Survey is carried out each month on a random sample of about 1,800 adults, aged 16 and over, living in private households throughout Great Britain. After accounting for refusals and where no contact can be made, approximately 1,200 interviews are conducted each month.' It doesn't say if this survey is conducted online, but presumably it isn't ...

proportions or percentages it's worth asking what the underlying numbers are.

In this case we don't need actual values, because the proportion is so obviously grotesque. It is saying that 20 per cent of men who live to be 65 – and the majority of men do live to be 65 – will die within two years of making their 65th birthday. That would be a terrible mortality rate. What the statistics actually *said* was that around 1 in 5 men will die some time before their 67th birthday. But most of these won't die in that two-year window between the ages of 65 and 67.

There are, certainly, plenty of statistics about dying – it's a topic that inevitably interests us. As we've seen (page 54) some of the earliest work in statistics involved life expectancy, and there is a rather bizarre unit of the risk of dying called the micromort. This is a way of looking at risk – a micromort says that you have a one in a million chance of dying. If you look at the average rate of death across the population and divide by the number of days in the average life, you can expect about 40 micromorts a day overall as the chance of randomly dying – but of course this ignores anything that takes you away from the average, say by being very old or indulging in risky activity.

Micromorts are sometimes used to compare the risk of different activities. It is possible to say, for instance, that you increase your risk of dying by one micromort by smoking 1.4 cigarettes, travelling around 250 miles by car, flying 1,000 miles by plane or having a chest X-ray. (If you want a really risky means of travel, a micromort takes just 6 miles on a motorbike). This is an amusing exercise, but it has to be borne in mind that all you are doing is getting a rough statistical impact, not a detailed assessment of your personal risk. Apart from anything else, not every risk is linear – you can't assume that smoking 40 cigarettes a day is exactly 40 times as risky as smoking one cigarette a day. But micromorts, if used properly, can give us some insights into the comparative risk of activities.

Blinded by science

Sometimes statistics can be so confusing that a professional making use of them can be just as easily misled as can the layperson. This is very easy with the statistics of a medical test. Let's imagine you take a test for dangerous disease, a more and more frequently used part of the clinician's diagnostic toolkit. Say the test is 95 per cent accurate, by which we mean that 95 per cent of the times the test is given, if it says that you have the disease, that test is correct. You really are ill. But in 5 per cent of tests, it will say you have the disease when you don't. You take the test and the results come back positive. It says you have the disease. What are the chances you are ill?

It seems ridiculously simple. Surely there's a 95 per cent chance you have the disease? And this is a conclusion doctors have been known to come up with. But it's horribly wrong. In fact, given the information above I can't say how likely it is that you are ill. That's because I don't have enough data. Let's say the test is given to a million people a year, of whom 1 in 2,000 have the disease. That means 500 of the people who were tested will have the illness.

So 500 of the positive test results will be genuine*. How many positive test results will be wrong? That will be 5 per cent of the total taken – 5 per cent of one million, which is 50,000. So there will be a total of 50,500 people with a positive result, of whom 50,000 aren't ill. If you have a positive result then there's only a 500 in 50,500 chance that it's genuine – just under a 1 per cent chance that you are actually ill.

An even simpler problem of comprehension is getting your head around numbers that you just don't come across in everyday life. Governments are always bandying around numbers in the millions and billions of pounds or dollars. (When it comes to national debt they can even throw trillions around.) This is outside our everyday experience, so such numbers really don't mean a lot to us. C. Northcote Parkinson was the historian who came up with Parkinson's Law, which states that work expands to fill the time available for its completion. As he comments, 'General recognition of this fact is shown in the proverbial phrase "it is the busiest man who has time to spare."' But he also had a useful observation about the way we deal with large numbers.

Parkinson imagined a meeting among non-experts (politicians, say) discussing the construction of a new nuclear power station. They would, he suggested, spend very little time on the big thing, the nuclear plant itself, as none of them understood it. It involved numbers, science

---

* I'm simplifying here by ignoring the fact that most real tests will fail to spot people who really do have the disease in some cases, so in practice you would have slightly fewer than 500 positive test results for genuine cases.

and engineering on a scale that was totally outside their comprehension. But they would spend an inordinate amount of time on discussing the siting of the bike sheds, as this was more on the scale they understood. It's the same when we think of large numbers in statistics.

Imagine a US President announcing, 'Next year we will need to raise a total of $1.1 billion for a new initiative that will transform lives.' Let's not worry what the initiative is, but imagine it is clearly beneficial. It sounds like a huge amount of money. Now let's imagine you wanted something and it cost one cent per person per day. Comparing that with, say, the cost of a coffee, that sounds a very reasonable investment. But both are, in fact, the same number. There are around 312 million people in the United States at the time of writing. That means a cent per person per day is around $1.1 billion, while a dollar per person per day adds up to $113 billion. Large numbers only make sense to us when taken to the personal level.

All in the context

We need to be particularly suspicious of statistics that are based purely on percentages, because without context the percentage is practically meaningless. Imagine a headline screaming 'Murder rate in city increased by 100 per cent!' It would probably encourage you to stay in at night if you were visiting that city and not risk going out onto the mean streets where, in all probability, you would be butchered. But what if the previous year's murder rate had been one victim? Then a 100 per cent increase is one extra killing. Of course any murder is terrible – but there will occasionally be some, and the fact there is one more in a particular year is neither here nor there – and is certainly not a good reason for hiding in your hotel room.

Here's another example you will often see in those interminable media stories of how eating and drinking different things can have negative or positive impacts on your health. Let's imagine (and this is purely imaginary) that eating oranges was shown to increase your risk of getting a particular serious disease by 50 per cent. It sounds scary. Consumption of oranges would probably collapse overnight when the news broke. But with only a percentage to go on, we know practically nothing. If the original risk was 0.01 per cent – so around one in 10,000 people would get the disease – then we are now saying that 1.5 in 10,000 (or one in 6,666) are at risk. It is still a very unlikely occurrence. That extra information is essential in judging the significance of the statistic, and whether or not it should influence your consumption of oranges.

One more example of using a single figure in a way that is downright confusing. A while ago, a spokesperson for the Soil Association, the main British organic accreditation body, made a dramatic claim in a newspaper. 'You can switch to organic,' she said. 'Or you could just accept that every third mouthful of food you eat contains poison. Are you up for that?' Scary stuff. But the statistic was problematic.

Firstly, that 'every third' figure was far too low. In reality pretty well every mouthful of food you eat contains poisons. Some are the pesticides and other manmade chemicals she was referring to, but many types of food also contain natural poisons, aimed at protecting plants from damage. Some of the most deadly poisons – ricin and the botulinus toxin, for example – are natural. And there are also many natural substances in food that have been shown to produce a cancer risk for anyone consuming them. So the fact that the Soil Association was basing its statistic on – that around 40 per cent of fresh fruit and vegetables retain traces of agro-chemicals – was beside the point.

However, even if you don't worry too much about this distinction, the percentage did not provide enough information to be useful. Because toxicity is all about dosage. Practically everything is poisonous if you consume enough of it. Water, for instance, can kill you, if you drink too much. I'm not talking about drowning – athletes have died after drinking several litres at a time because the water reduces the effectiveness of signalling in the nervous system, resulting in the electrochemical messages that control the body failing.

In that famous book that almost single-handedly started the environmental movement, *Silent Spring*, Rachel Carson made a strong point. 'For the first time in the history of the world,' she wrote, 'every human being is now subject to contact with dangerous chemicals, from the moment of conception to death.' This too was wrong. Throughout all of history human beings have been subject to natural poisons. We typically consume around 10,000 times as much of deadly *natural* pesticides as we do of artificial ones. But even this is a very small quantity that will do no harm.

The fact is, if you look at the risk from the quantities of poisons we consume, the last thing we need to worry about is pesticide residues. Looking, for instance, at the cancer risk from the average diet, 93 per cent of the risk comes from alcohol, and 2.6 per cent from coffee. Once we get the relatively dangerous natural sources of risk like lettuce, pepper, carrots, cinnamon and orange juice out of the way, the first chemical contaminant is a chemical called ETU at 0.05 per cent. If you add up all the major chemical contaminants and pesticides at legal levels, they have a similar risk to eating celery.*

Statistics aren't fair

Another issue we have with the whole business of randomness and statistics is a rather more fuzzy one of the impact on feelings. Statistics can be horribly cruel. Think of the poor average person. On average, a person has fewer than two legs, fewer than two functional eyes – in fact the average person doesn't have a full set of anything that you can lose and still survive. The vast majority of drivers are of the opinion that they are above average ability. Yet the majority must be average or worse.

The distribution of driving ability is probably roughly symmetrical, and in the case of both driving skill and, for example, human height, there is only so far you can

---

* I must stress again that the risk from consuming the likes of lettuce, pepper, carrots, cinnamon, orange and celery is very low. They all have a tiny risk of giving us cancer, but it is so small that it can comfortably be ignored. Practically everything you eat has some risk attached to it. The point here, though, is that the risk from pesticide residues is even lower.

go to either extreme. We don't have 10ft-high people, for example. But we have to bear in mind that not every distribution is like this. Let's imagine that we have a distribution of earnings or wealth in a room full of people that is roughly a normal distribution and relatively limited in spread, then we throw in a billionaire. Suddenly the average will shoot up, meaning the vast majority of the population will be far below the average level of earnings and wealth. When Bill Gates enters the room, *everyone* else has a below average net worth. That's not great for your self-esteem.

One average that has led to a widespread misunderstanding is historical life expectancy. We all know that before the 20th century people lived short, unpleasant lives. This is a quote from a recent book about the brain: 'My grandmother, passing her 88th birthday was unusual. Life expectancy for girls born at the beginning of the twentieth century was just 49 years, for boys 45.'* Unfortunately the author of that comment, like many others, fell into a statistical trap.

It is true that her grandmother making it to 88 was relatively unusual. But the life expectancy figures are totally misleading. Back then, some middle-aged people would have died due to poor sanitation and an inability to prevent some diseases that are now treatable. But the fact remains that, if you made it into your forties in the 19th century, you would probably live to your sixties or seventies. The reason the life expectancy figures

---

* From *The Brain Supremacy*, Kathleen Turner (Oxford University Press, 2012)

were so low was because so many people died in infancy and childhood. This pulled the average down very significantly. It wasn't that most people died in their forties – just that the average, with all those very low numbers, was dragged down to that level.

This distortion of the average is something that politicians wishing to attack another party and not being too scrupulous about their statistics can use to their advantage. If a tax is aimed at those on high earnings and it turns out to hit people on the average wage, then there is an outcry, because that seems to imply that it hits the majority of ordinary people – but the majority earn *less* than the average wage. The politician can play the numbers even more effectively by putting two people on an average wage into a household. Now we are not only using individuals that earn more than most as individuals, but a household where both partners do so. This pushes their collective income up so high that it puts the household in the top 25 per cent of all households, even though we are talking about two people who each earn the average wage.

It's pretty obvious that if Bill Gates came into a room of people on middling ordinary wages, then the outcome would be to heavily skew the average upwards. What's less obvious is why the population as a whole is skewed so that most people earn less than average. Taken across those in employment as a whole, 80 per cent of the world earns less than the average. Given that there are more poor people than there are rich people, why don't those poor people pull the average back down? The reason this doesn't happen is that the measure we use (earnings, say, or net worth – the total value of all your property, money, shares etc.) has a bottom limit but not an upper limit.

Generally speaking we can't earn a negative amount or have a negative net worth. (In the short term you can if you have less income than your debts, but you will rapidly go bankrupt, resetting the value to zero). At the time of writing, the average annual income in the UK for those in work was around £26,000 and the equivalent in the US $46,000.* Someone on no income can therefore only be £26,000 or $46,000 below average. But a very wealthy person could easily earn £1 million more than the average, and so would have a much bigger impact than any single poor individual.

Taking an average can be very useful but it has to be handled with care. In the example of the room with Bill Gates in (if you want to spot him, he'll be the one everyone else is trying to put business proposals to), it would be more useful to ask what the median earnings or net worth is. Median is a heavy sounding word for a very simple concept. If you get a list of all the values in order, the median is the middle one. In a nice symmetrical distribution like a normal distribution, the median and the average value (the mean) will be pretty similar, but with Bill Gates skewing a distribution wildly towards the top

---

* Dealing with income statistics can be particularly tricky, because there are so many variables. Are we dealing with net income, after tax, or gross income before tax? The average income I gave above would be higher if we only included those in full-time work. It would be less, however, if we included the population who could work, but aren't necessarily working right now, and even less as a whole by including children and the retired. You pays for your statistic and you takes your choice.

end, the median will be well below the average, and will give a better idea of the *typical* value.

Official statistics often do make use of the median where it is more meaningful than the average, but newspapers and TV (and devious politicians) for some reason prefer to use averages. Where the average income for the UK in 2011 was £26,871, the median income was £21,326 – emphasising the difference between the average and a 'typical' earning. Because the media shy away from using the term 'median' (presumably because they feel it's too complicated for the ordinary reader), you will often see them reporting the median but calling it the average. The figures above are for all workers. The median income for full-time workers was £26,200, yet many of the newspapers reported that the *average* for full-time workers was around £26,000.

The rare event

The basis of statistics and sampling is the assumption that underlying the numbers is true classical randomness, and as we have seen, such random values typically fall into some kind of distribution, whether it is the familiar bell-shaped normal distribution or something more complex. This means we have to be very careful when handling a statistic that can include sudden wild values – as in the effect when Bill Gates enters a room and totally distorts the earnings or net worth of the group. It is fairly obvious what is happening with Gates, but it's harder to imagine immediately what the problem is if you hear that the safest type of airliner in the world has suddenly become the most dangerous. Have planes been plummeting from the sky all over the place? No, it took just one crash in 2000.

On 24 July 2000, Concorde was, statistically speaking, the safest plane in operation. The Concorde fleets had never had a single crash during their entire operational lives. But the next day, Air France flight AF4590 crashed as it took off from Charles de Gaul airport in Paris. One disaster. But there were few Concordes operating – Air France and British Airways had a handful each, the sum total of operational aircraft – and each plane made relatively few journeys a year. And so the risk immediately shot up after a single accident to around one crash in every 80,000 flights, compared with around one in 3 million for some of the other airliners.

Concorde hadn't changed. The risk hadn't changed. But when we're talking about such a rare event, the statistics can be highly misleading and we do need to know the frequency of the activity and how it compares with rivals if we are to make any sensible comparison. There is a distribution called the Poisson distribution, which looks a bit like a normal distribution that has been squashed up to one side, that is quite effective at giving us an understanding if the actual frequency of the event is reasonable given the expected average, or if the rate is such that there is likely to be a new factor involved.

As the frequency increases, the Poisson distribution becomes more symmetrical*. The Poisson distribution

---

* Note, by the way, that the Poisson distribution is always a distribution of individual ('discrete') events. It hasn't got a continuous value, so any graphic representation should show a series of dots rather than an uninterrupted curve.

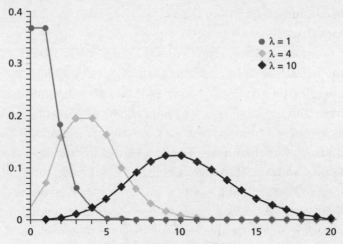

Note – λ is the average number of events in a time interval

A Poisson distribution

can be useful, for example, in understanding if a rare event that appears to be part of a cluster with a cause actually does have a cause, or whether taken across the population you would expect clusters of that size to occur randomly. But it won't distinguish usefully between a Concorde and a conventional airliner.

If we are comparing like with like, on a sensible scale, for values that occur at random on a known distribution, and those occurrences are independent of each other, we have real prediction power in our statistics. But there is an assumption underlying all this. We merrily speak of values picked at random, but what does 'at random' mean?

CHAPTER 9

What does random mean?

Let's imagine we had a special machine for generating random numbers. It doesn't really matter how it works – we can come on to that later. What would a true random number generator actually deliver? We'll limit ourselves for the moment to random whole numbers between 1 and 10 inclusive. Then our generator should give us a string of numbers, each between 1 and 10 in value. Every time a number is selected there should be a 1 in 10 (10 per cent) chance of any one of those numbers being chosen. And each choice would be entirely independent from the rest. The random number generator would have no 'memory', no link between the choice of a number and the choice of subsequent or past numbers.

That's clear. The output of such a generator is a true random number sequence. However we might find the results difficult to accept. Say I gave you the following sequences to check:

    1   1   1 ...
    1   2   3 ...
    5   1   4 ...
    6   9   2 ...

Let's imagine that these are the first three numbers that came out of our random number generator. Which of these sequences really are random? Which seem more likely to occur? Our natural inclination is to say that the third and fourth sequences are the sort of thing that we would expect from the random number generator, while the first two aren't. This is a rather strange assertion, as it manages to be both true and untrue at the same time.

The first thing to realise is that it is perfectly possible for a true, working random number generator to come up with a sequence like 1 1 1 or 1 2 3. Remember, one of the essential rules of our generator is that it has no memory. Whatever came before a number will not influence the future. The chance of getting a 1 as the first value is 1 in 10 – just like the chance of getting any other number. We certainly can't deduce anything from getting a 1. Then we get to the next number. Because the generator doesn't know that the first number was a 1, once again, the chance of getting a 1 is 1 in 10.

Combining these chances, the chance of getting the sequence 1 1 is 1 in 100. The same logic applies to the third value. Here again there's a 1 in 10 chance of getting a 1. So the overall probability of getting the run of 1 1 1 is 1 in 1,000. It's a low probability, but perfectly feasible if you have enough attempts. Think about it – the chance of any particular set of numbers being drawn in the main UK National Lottery (Lotto) is 1 in 13,983,815 – but a set of values does get drawn every week. Exactly the same reasoning goes for 1 2 3 as 1 1 1 – your chances of getting this too is 1 in 1,000, so if you run the generator thousands of times you would expect this to come up.

The interesting thing, though, is that the same probability also applies to any sequence of three numbers, including apparently more random values like 5 1 4.

This sequence too has a 1 in 1,000 chance of turning up. So none of these sequences is more likely to be produced than any other. Each of them has exactly the same chance of being selected. There is a book, sometimes labelled 'the most boring book ever', produced in the 1950s, that lists 1 million random digits. In that sequence you are likely to find around 100 examples of five sequential digits in a row, around 50 examples of the same number repeated five times and even one example of the same digit repeated seven times.

Why is it, then, that we consider a sequence like 1 1 1 to be surprising and challenging to our idea of randomness? Of course it's the brain's pattern-recognition software at work again. And it is understandable, even though our suspicion isn't based on pure probability. While 1 1 1 is just as likely to come up as any other set of numbers, the *kind* of sequence – one with a clear visible pattern – is much less likely. If you think of the different ways to arrange a sequence of three numbers between 1 and 10, only a few of them are likely to stand out as special.

There are 1,000 possible arrangements in total. We are likely to think that a sequence has a pattern if all the numbers are the same – so that gives us ten sequences from 1 1 1 to 10 10 10. We will also spot it if the numbers make a nice, evenly varying sequence like 1 2 3 or 4 5 6, or for that matter, 10 9 8. There are 16 of those. But most other sequences wouldn't seem strange to us. (You could argue for sequences with some kind of regularity like 1 2 4 or 1 5 9, but many people wouldn't label those as special.) So we have 26 special sequences that stand out to us and 974 not special sequences that don't have features that catch our attention.

It's not surprising, then, that we consider the clear and obvious patterns as being special cases. They aren't, but we can see why they look strange. We have to be on our guard not to treat them any differently unless there is other evidence. Simply getting the sequence 1 1 1 does not mean that we have a faulty random number generator. Any specific section of a sequence of random numbers can contain a clear pattern and with a long enough sequence some of them will. The danger is, though, that because we are so focused on patterns as human beings, we will single out a pattern within a longer sequence, even though it is the very process of singling out that prevents us from having true random numbers, not the pattern itself.

If you ask a person to imagine a list of random numbers, you can be sure that their list will not contain enough repeated values. This has been observed time and time again. Because we tend to think of random as being another way of saying 'without a pattern' it feels wrong to include repeated values – and it looks wrong when they occur. But without some repeating numbers a sequence cannot be truly random. This is the same effect as clustering. If all the numbers in a sequence are well separated from each other it is just as unnatural as a can of ball bearings dropped on the floor producing a set of ball bearings that are all carefully spaced away from each other.

This has proved a problem in the past when attempting to test for telepathy experimentally. In some trials, a random number generator selected a number when a human operator (a research student) pressed a button.

Then the operator would attempt to communicate that number (or a playing card corresponding to it) by telepathy. If the generator produced the same number twice in a row, the operator had a tendency to assume that they hadn't pressed the button properly and pressed it again. This simple error corrupted the experiment and made it seem as if people were telepathic even though there was no such evidence from this experiment.

How could this happen? There were ten possible numbers. So if the person being tested for telepathy guessed at random, with no telepathic effect, they should have a 1 in 10 chance of getting the right answer. If they scored better than 1 in 10 over a long series of trials then there must be something happening, perhaps telepathy. But with the operator rejecting a repeated number, there was a 1 in 9 chance of the number that was used being chosen. And because the people making guesses *also* tend to avoid repeated numbers, they will have a 1 in 9 success rate. And that's enough better than 1 in 10 that over a large number of trials it could be assumed that telepathy really exists.

Experiments that try to prove or disprove the existence of telepathy and other psi phenomena are bedevilled with errors arising from probability and statistics that can produce an apparently significant result from either failing to use truly random numbers or misinterpreting the results.

Generating randomness

It's easy to understand how there might have been difficulties getting truly random numbers before the computer age. Anything physical like coin-tossing has the potential for bias either in the way the action is undertaken or in subtle influences. As we have seen, because of the way we flip coins, a coin is slightly more likely to end up showing the same face when it comes to rest as was facing upwards initially. So really, to keep a sequence of coin tosses fair, you should alternate between starting heads up and starting tails up.

One approach that was taken in the early days was to turn tables of logarithms into crude random number generators. These mathematical tables, which were used for multiplication and division before calculators, contain a large number of decimal place values, and were widely available. By using a fixed set of instructions, for example, 'Take the seventh digit of the 93rd number', it was possible to generate quite a good set of near-random values. But log tables were usurped by calculators and then by computers. And the obvious question is why the random number problem didn't go away when computers were introduced. After all, if I fire up Excel or any other spreadsheet I can generate a random number with ease.

In Excel I have two random number functions, RAND, which gives me a random value between 0 and 1 (I just got 0.61012053) and RANDBETWEEN to choose a random number in a range. (My value for between 1 and 10 came out as 4.) Job done. Unfortunately, what Excel gives us is not random numbers, but pseudo-random numbers. Numbers that are random enough for, say, making a prize draw, but that aren't good enough if you want a good long sequence of numbers that are truly random.

That is because the pseudo-random number generator is not genuinely picking between the options with
equal probabilities, nor is any value in a sequence of
numbers it generates totally independent of what came
before. That has to be the case or we can't calculate the
random number using some sort of computer algorithm.
A spreadsheet's pseudo-random generator usually starts
with a 'seed', an initial value which is often taken from
the computer's clock, and then repeatedly carries out a
mathematical operation, typically multiplying the previous value by a constant, adding another constant and
then finding the remainder when dividing by a third constant. So, for instance, a crude pseudo-random number
generator would be something like:

$$\text{New value} = (1{,}366 \times \text{Previous Value} + 150{,}899)$$
$$\text{modulo } 714{,}025$$

where 'modulo' is the fancy term for 'take the remainder when you divide it by ...' The output of the pseudo-
random number generator wanders off away from the
seed value and can be reasonably convincing in appearance, but it will always produce the same values given
the same seed, and can't match a true random number
generator for effectiveness of results. Although my Excel
RANDBETWEEN function can produce two repeated values, because it is rounding a wide range of real number
values to get to the same figure, the pseudo-random generator will still always be limited because it can never
produce the same *exact* value twice in a row or it would
get locked into repeating that value over and over.

Those who want to be more careful about their randomness look for a better way of producing their outputs. Most large lotteries rely on machines with a series of balls in them, which are randomised by stirring them up before balls are drawn. This is not the best way of getting a random number by any means. It's still a pseudo-random number generator. The chances of the balls being drawn in a perfectly random fashion are very low. But in this particular example, visibility is more important than perfection. It is considered more important for players to see the draw happening than it is to approach the perfection of true randomness.

This isn't the case for all lotteries, though. In the UK, for instance, we have a kind of lottery known as premium bonds. These are government bonds that allow the buyer to have a little flutter, instead of providing a predictable return for all purchasers, as is traditional with bonds. Most premium bonds will return no 'interest' (though unlike a lottery ticket they can be cashed in to get the initial stake back). Only 1 in 24,000 of the £1 units wins a prize in each draw. But some bonds will produce this cash return, which can vary from a few pounds to £1 million.

To make the draw fair, the people behind premium bonds were among the first in using electronic random number generation with a device known as ERNIE, a contraction of Electronic Random Number Indicator Equipment. This device was introduced in 1957, designed by Tommy Flowers, the man behind the world's first electronic programmable computer, the Colossus, at the World War II code-cracking centre, Bletchley Park.

ERNIE worked by using the noise in the signal produced by a series of valves (vacuum tubes). All electronic devices produce a degree of noise due to thermal variations in the materials, interference and other effects. Although not truly random, because the outcome could in principle be predicted if you had all the available data, no one has access to that data and the effect is sufficiently chaotic that it is impossible to have any idea what will come next. Because of this, it is a safer way to generate pseudo-random numbers than a software-based approach. More recent variants of ERNIE (they are on to Mark IV) make use of thermal noise in transistors.

True random values, as we will see in the next chapter, can be produced using quantum effects, and true random generators are now available to plug into electronic devices if required. For example, if you have a radioactive source, where atoms occasionally spontaneously emit part of their nucleus, you can predict how often an atom will undergo such a decay on average, but the decay of a specific atom is truly random: it is not just impractical to predict but is impossible because there literally is no cause. The modern-day ERNIE could be based on such a system, but the approach taken is equally good in terms of being unpredictable and is easier to produce.

Cherry picking

No matter how good your random number generation, it is still possible to influence the outcome of a trial based on probability. One way to do this, either intentionally or accidentally, ruining the value of the experiment, is to be selective about your results. A crude approach (but one many researchers are suspected of) is simply to ignore any trial where the results don't match what you want them to be. But selectivity can be much more subtle. Let's imagine I'm doing a trial of something that is randomly distributed, but that I assume is due to some special ability – say I am testing people for the ability to toss better than the average number of heads with ten tosses of a coin.

I get a large group of people and ask each to toss a coin ten times. Most will get around five heads. But we would expect a few to only achieve 1 or 2 heads, and a few to get 8 or 9 heads. All I do then to bias the outcome is split off my 'experts' and only use their scores. Here I have a group of people who are most likely to be able to mentally influence the tosses of coins because they got the best results. I can even ask them to do some more tosses. The chances are they will only achieve around a 50 per cent success rate with these new attempts, but as long as I still count the original scores, overall their performance will remain well above average. The act of selection from a group distorts the significance of their scores.

There is no problem with doing a test run of an experiment like this to pick out those who could have the appropriate ability (should it exist) – but it is essential to then discard the values used for selection and start testing again from scratch. However, because of limited time and money it is always extremely tempting to include the data from the selection run, and immediately bias has been introduced.

Another way to misuse data is to pick out significant *sections* of runs and ignore the rest of the data, so-called cherry picking. Let's imagine I am doing a coin-flipping experiment with 1,000 flips. Part way through my sequence I get seven heads in a row. Over the whole experiment, everything pans out nicely – let's say I get 495 heads and 505 tails. Nothing special. But if I just home in on those seven heads, I can say the chances of this happening is $\frac{1}{2} \times \frac{1}{2} \times \frac{1}{2} \times \frac{1}{2} \times \frac{1}{2} \times \frac{1}{2} \times \frac{1}{2}$ – just one in 128. Pretty unlikely. In truth there was nothing special about that sequence given the number of tosses overall, but if I search for it and pull it out of the data, I give it an unreal significance by making that selection.

That one in 128 chance I just came up with doesn't give the impression of just how seductive this kind of reasoning can be. A better example comes from the work of J.B. Rhine on telepathy in the 1930s. Rhine did many thousands of experiments using packs of cards, each card having one of five possible designs, to test for telepathy. In one run out of those many thousands, the person guessing values got fifteen correct answers in a row. As it happens, Rhine's experiments have since been criticised for having insufficient controls to prevent cheating, so we don't know if this run was genuine. But the way Rhine reacted to it certainly lacks scientific credibility.

He comments on this 'brilliant run of 15 unbroken hits' using some exquisite cherry-picking of the statistics. He says 'The probability of getting 15 straight successes on these cards is $(\frac{1}{5})^{15}$, which is one over 30 billion.' It sounds impossible that this could happen by chance alone. Clearly there has to be a cause. But Rhine has picked out the sequence from those many thousands of guesses *because* it is outstanding. Put it into context of the full set of data and it has very little significance. It is only by cherry-picking it that it looks so amazing.

An even more subtle cherry-picking error that is particularly common is to give the researcher the chance to discard the results if things go wrong. Imagine, for example, you were measuring how well someone scored at an activity that required a lot of concentration. Part way through the trial there is a loud bang and everyone is distracted. Someone has crashed a car outside. After the trial, the scores are low. So you discard the results because an outside factor influenced the outcome. That is fair enough. But if the scores had not been low, you would have probably kept them on the assumption that the distraction did not cause a problem.

What seems harmless and sensible has distorted the outcome. What you have just done, probably unwittingly, is to discard 'bad' results and keep 'good' results, where bad and good are defined by the kind of outcome you want from the experiment. This introduced bias does not mean you can never discard results – but you need to decide in advance in what circumstances you will do so, and should make the discard without knowing what the results are. So it would be perfectly legitimate to say 'if there is a loud noise, we will discard the results and not look at them'. But not 'if there is a loud noise *and* the results are poor we will discard them.' That second type of rule is often applied without us realising it is being used.

The unbalanced target
Anyone designing experiments that depend on probability has to take a huge amount of care, and would do well

to bring in a probability expert to check that there are no pitfalls. Even the way that the data is collected can influence the outcome. This seems counterintuitive, like so much of probability, and you have to be unlucky to hit on one of the perils, but it can happen, particularly if the phenomenon you are studying depends on how a random sequence arises.

Here's a simplified model of data-collection in action that demonstrates how randomness can trip you up. We'll do a mock experiment where the data being collected is simply the value of a tossed coin – heads (H) or tails (T). We are looking out for a particular pattern in the data, and will stop the experiment as soon as that pattern appears. We will run this in two ways: firstly where the sequence H T T is the trigger for stopping the experiment, and secondly where the sequence H T H is the target. The crucial piece of data is then how many coin tosses are necessary before we get to our target.

As is usual with this kind of experiment, we repeat the process over and over again to get a statistical picture. The important question is, would you expect, on average, that it would take more coin flips to reach the H T T target, more to reach the H T H target, or the same number of flips in either case?

It is pretty obvious that each target is equally likely to come up – the chances of getting each sequence is 1 in 8. So if you ran the experiment, or a more complex one using the same kind of reasoning, and it turned out that, say, it is taking fewer flips to get to H T T than it is to get to H T H, you would suspect there was something wrong with the experimental design. Or worse still, that someone involved in the experiment is cheating. They must have something to gain from making the outcome different from expectation.

The only thing is that this result will happen with no one cheating. In such an experiment, H T T will, on average, take fewer flips to reach than will H T H.

To see why, think through the process of getting to the desired result. In both cases, you will first need to have tossed H then T. Then imagine what happens next in each case, starting with having a target of H T T. There is a 50 per cent chance I will flip a tail, and I've reached my goal. And there is a 50 per cent – 1 in 2 – chance of getting a head, in which case I have to carry on. In this case I have just got 'H', so now to succeed I have to toss T T. Each has a 1 in 2 chance. So if I do fail on my original attempt, I then have a 1 in 4 chance of succeeding with two more throws.

Now think of what happens with a target of H T H after already tossing H T. Again there's a 50:50 chance of getting H or T on the next toss. If it's H, the target is reached. If not I have a T. Now here's the interesting bit. Remember in the previous case at this point I had a 1 in 4 chance of succeeding with two more throws. But in this case I have a 0 in 4 chance of doing so. Why? Because the first toss of my sequence is T. I can't produce H T H *whatever* the next two tosses are. It is only once I have cleared that leading T that I can carry on with any hope.

How significant is significant?

The situation described above shows how it is possible to trip over the maths. The other danger when using a probabilistic assessment is tripping over the logic. This is more likely to happen in casual use, but even scientists who are making use of a statistical approach have been known to do this. This pitfall occurs when we are making an observation and use statistics to show that (for instance) there is only a '5 per cent probability of this result happening by chance alone.' Let's see first where that '5 per cent probability' came from, and then look at the reasons why it is very easy to get the implications in a twist.

What we are dealing with here is a 'significance level', often abbreviated by scientists as sigma (σ). The starting point is having a distribution. Let's imagine we are dealing with the kind of random event that fits a normal distribution, producing a bell-shaped curve. The most likely results are in the middle of the curve. As we move towards the edges we get two thin 'tails' heading off to infinity in either direction. These are the less likely events.

As we have seen, quite a lot of simple real-world random events, ones obeying classical randomness, fit a normal distribution, though many more will follow a different distribution or be chaotically random. A commonly used example of a natural-world normal distribution is the spread of human heights – this is a great example for understanding what we're dealing with, especially because in reality human heights don't fit a normal distribution at all.

To start with we need to be a little careful about just what we are measuring. It's probably best to stick to a single sex, as clearly we would otherwise be combining two distributions, as average male height is significantly greater than average female height. Let's say we just look at males. Then you would need to be careful to take your sample (because you are unlikely to have measurements for everyone in the world) across a good cross-section of the population, rather than just in a shop selling clothes for tall people (say). But even then, if you look at a distribution of heights, you might get a surprise. Even with these restrictions, it won't be normal.

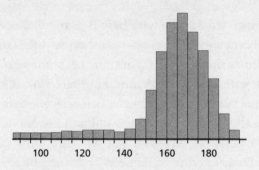

Male height distribution (in centimetres)

Using data for US men, you might be surprised to learn that the average height is just 5ft 6in (167cm). This is because, rather than being a symmetrical distribution, the range of heights has a much greater spread at the short end than at the tall end. Around 99 per cent of all men are no more than 11in (28cm) taller than the average height. But to get to that 99 per cent figure you have to include men as short as 2ft 7in (79cm) shorter than the average height. If you look at a graph of the distribution, the tail-off to the left – the short men – is much longer than that to the right. This also shows in the way that the median height, the middle value, is 5ft 8in (172cm), significantly taller than the average. Most men are above average height.

As we have seen, we also have to be careful with applying these distributions where there is a chaotic element to the randomness. If we look at net worth or book sales, a few individuals can hugely distort the whole picture, making a traditional distribution almost worthless. This isn't the case with the height distribution. There are clear limits. You will never find a man with twice the average height, whereas someone with twice the average earnings is fairly commonplace. And even though height does not fit a normal distribution, the distribution of heights can be dealt with well in determining a significance level.

For the purposes of understanding significance levels, let's assume that we are looking at something that does have a normal distribution – a nice, symmetrical bell-shaped curve. If we just take the tall middle section and cut off the two shallow tails we will capture most of the values. The 95 per cent confidence level is the central chunk of the distribution where 95 per cent of the values are expected to fall. So should we get a result that falls in one of the two tails that are left behind when we take out that middle chunk, it will only happen by chance 5 per cent of the time.

The measures of confidence in a result are often expressed in terms of 'sigma' levels. Sigma is the symbol for 'standard deviation', which is a measure of how quickly a distribution spreads out into its tails. The higher the number of sigmas, the less likely it is that the result would be obtained purely randomly. Our 95 per

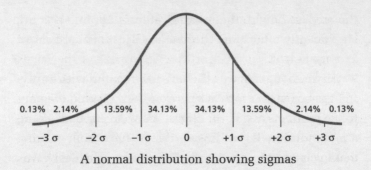

A normal distribution showing sigmas

cent confidence level is a 2 sigma level. As we have seen, when there were indications that fitted with the Higgs boson from the Large Hadron Collider at CERN in 2012, this was reported to be to a '5 sigma level'. This was the equivalent of saying there is just a 1 in 3.5 million probability of these results happening anyway without something that looked like a Higgs boson.

So given an appropriate distribution of observations of something that is classically random we can put a figure to how likely it is that the results observed would occur at random with no cause. Here comes the pitfall. Let's say we are trying to test a hypothesis like 'the Higgs boson exists' or 'these results were produced by telepathy' or 'this cluster of cancer cases was caused by that phone mast'. What it is very easy to do is to say that as there is a very low probability of the observed result occurring by chance, there is a high probability that the hypothesis is true. This is *not* what the statistics imply.

By showing there is a low probability that this observation was down to chance alone, all we prove is that there is a high probability that it has a cause. Any cause. Certainly not necessarily your hypothesis. So, for example, when we say that the Higgs results showed a 1 in 3.5 million probability of occurring if the Higgs boson did not exist, this did not mean that the cause was a Higgs boson, but rather a particle that behaved the way we expect (one type of) Higgs boson to behave. Some media sources complicated things further by saying there was only a 1 in 3.5 million chance that there was no Higgs boson. This is subtly different from the truth, which is that there is a 1 in 3.5 million chance that these data would be observed with no Higgs boson to cause them. The incorrect version focuses on the Higgs (or its absence), but really the probability is about the data.

A similar trap of assuming that the unlikeliness of one possibility proves the likeliness of an alternative is a favourite pitfall for supporters of the concept of intelligent design (ID) in biology. Though many think that intelligent design is just creationism with a scientific gloss, ID supporters argue that they have no axe to grind on who or what the designer is, they merely want to show that there has to be active design at play, rather than the random influence of evolution.

A typical approach they take to support intelligent design is to look for structures in nature which they believe could not have evolved because there was no benefit to the organism from an intermediate step. So, for example, they point to the biological 'motor' used to propel some single-celled organisms' propeller-like flagellum. The individual components of this motor, they suggest, have no value, so the usual evolutionary argument of gradually changing in a beneficial way could not apply. So, they argue, ID is correct. They have shown that gradual evolution through incremental beneficial changes has a low probability of producing the result – and so decide their hypothesis is correct.

However, even if it is true that there is no benefit from the intermediate stages (there often proves to be in practice), all the ID fans have actually shown is that this feature is unlikely to have developed as a result of that particular mechanism of evolution. They have not shown it isn't because of a different evolutionary mechanism, or for that matter for a totally different reason. ID is just one of the possible options that are still feasible – by disproving one hypothesis you don't automatically prove a second.

Probability on trial

If those arguing for a particular scientific theory get it wrong, then the only real risk is that false conclusions are drawn in an academic paper. But other misunderstandings of the nature of randomness can have much more dire consequences, particularly when probability is wielded in the law courts by those who aren't equipped to handle the numbers. The most infamous case was that of the trial of Sally Clark and the 'expert' witness evidence of paediatrician Professor Sir Roy Meadow. Meadow made use (or, rather, misuse) of statistics to condemn Clark from the witness box.

Two of Clark's children had died, both when under three months. This could have happened as a result of natural causes, described as 'Sudden Infant Death Syndrome' or SIDS, but the second child's death had made the authorities suspicious. A government study that was recent at the time of the trial in 1999 claimed that the chances of a child dying of SIDS with no contributory factors like smoking in the household was 1 in 8,543. Meadow claimed in the trial that the fact that two of Clark's children had died meant it was necessary to multiply the probabilities together, making an overall probability of 1 in 73 million. He claimed this would only happen by chance once in every 100 years.

In an attempt to put this difficult-to-handle number into terms he felt the jury would understand, Meadow said it was the equivalent of backing an 80 to 1 outsider at the Grand National four years in a row and winning each time. It's hard to know where to start in pulling apart just what went wrong with the use of probability here. Meadow certainly clouded the issue by comparing it to a bookmaker's betting odds, which are not actual probabilities but forecasts. But there are two huge problems with that 1 in 73 million figure.

The first issue is that you could only say that the chance of two SIDS deaths in the same family was so low if the two events were not connected, and so each could be treated independently. If, like throwing dice or a coin flip, there was no 'memory' in the event. However it is much more likely medically – something it seems very strange that Meadow appeared not to know – that once a parent has given birth to one child with SIDS they will also have another. If, for instance, there was a hereditary or environmental cause, you aren't starting with a clean sheet each time. A *British Medical Journal* report published after the trial suggested such events would occur not once every 100 years, but once every eighteen months in England alone.

The second problem is that there was a huge logical failure in progressing from the information that something had a low probability to the implication that somebody had acted in a nefarious way to cause that low-probability event. (Like the Higgs, this is a probability of the event happening, not of the hypothesis being true.) As we saw earlier, there is a 1 in 13,983,815 chance of a ticket winning the UK National Lottery's main draw, while the chance of winning the Euromillions draw is 1 in 116,531,799 – either side of that 1 in 73 million number. Yet week after week people scoop the jackpot.

It is very unlikely that someone will win – but when they do, we don't deduce that they have cheated in some way just because the outcome is extremely unlikely. The same should apply in the use of probability in court. Even if the chance had been just 1 in 73 million of both deaths occurring (which it wasn't), this doesn't mean there's a 72,999,999 in 73 million chance that it was a murder. You would have to compare the probability of this happening randomly with the probability that a mother would commit a child murder twice, which arguably would be less than 1 in 73 million. There was just no sensible comparison to be made with the data that were available.

The sources of randomness

As we have already seen, there are broadly two types of randomness. In classical randomness – perhaps representing the outcome of a series of coin tosses – we don't know ahead of time what an individual component's outcome can be, but we can put together a distribution that enables us to predict probabilities of different outcomes. In chaotic randomness, we could in theory predict an outcome exactly, but in practice we get caught out time and again because the complexity of the system means that shockingly large variations from anything that has happened previously can occur.

In the examples we have met so far, though, the inability to make predictions is always a result of not having good enough data and sophisticated enough models. The coin may have a 50:50 chance of coming up heads and tails in all throws, but on one particular throw, if I had every single bit of data about the environment, the coin and the way it was thrown, I should in theory, with Newton's help, be able to predict the outcome exactly.

Similarly, even though it is practically impossible to model the kind of system involved in deciding which book will be a bestseller and which won't, I can imagine a theoretical model that was as complex as the real world, modelling every person, every book and every decision, that would give me a good forecast. I could never do this practically because I could neither gather the right data nor have a complex enough model. (And because this is chaotic, a very small change in initial conditions could make a big change in outcome.) But in principle, with a computer that was as complicated as the world, I could crunch the numbers and tell you what the next bestseller will be.

However the real world continues to surprise us in the depth of its randomness. Because once we start to examine what is happening at the level of the individual particles that make up all matter, provide light and carry the forces that hold the universe together, we discover that there lies true randomness. Even perfect information and infinite computing power would not enable you to predict the outcome of a single quantum event. This is where true randomness rules.

CHAPTER 10

Really random

The apparent randomness driving the weather shows us the paradoxical nature of chaotic randomness – technically the process is still deterministic, but it is practically impossible to calculate exactly and, taken across many observations, less predictable than a distribution driven by classical randomness. Yet, in a sense, chaotic randomness is at least something with which we can feel comfortable.

It may be that we can't predict what is going to happen, yet we can rest happy on the understanding that underlying all the wildness is the familiar mechanics of Newton's clockwork universe. The interactions of all the components of the system may be far too complex to follow, but we can be reassured that any particular component will behave sanely and predictably. But 20th-century physics slipped a joker into the pack. It turned out that there was such a thing as true randomness. There is a monster that lurks at the heart of reality.

It all began with an attempt to patch up a theory that predicted the impossible. When you heat something up it begins to glow. Think of a piece of iron in a forge, being made hotter and hotter. To begin with there would be an infra-red 'glow' that we can feel with our skin, but not see. Hotter still, the iron will start to glow red, then yellow, then white. As it gets hotter and hotter it gives off higher and higher frequencies of light. But calculations made at the end of the 19th century predicted that something much more dramatic should happen – so dramatic, it was given the name 'the ultraviolet catastrophe'.

According to the best theories of the day, a blackbody (which is an imaginary object that emits or absorbs any frequency of light) should blast out radiation on all frequencies. The higher the frequency of light, the more energy should be emitted, adding up to an infinite burst of energy pouring from all blackbodies. Though a blackbody is a theoretical object, there are plenty of real-world things that come close to a blackbody – close enough to expect them to be pouring out a near-infinite torrent of light. Yet they don't. If it were true, we wouldn't exactly live in a stable universe. We wouldn't be here at all. There was clearly something wrong with the theory.

The solution to this catastrophe came from an attempt to fudge things. Max Planck, a young German scientist, found a way to contain the extravagances of blackbodies. Planck realised that the tendency to produce infinite quantities of energy only arose because it was possible to add up increasingly small increments of light waves at higher and higher energies. It's a bit like adding up the total of the series

$$1 + \tfrac{1}{2} + \tfrac{1}{3} + \tfrac{1}{4} + \tfrac{1}{5} \ldots$$

Even though each term in the series is smaller than the previous one, the numbers don't shrink fast enough to disappear away and the sum of all those fractions is infinite. What Planck realised was that you could bring the whole thing to a halt by insisting that an object could only radiate light in finite chunks. Imagine, for instance, that series of fractions was limited to chunks of $\frac{1}{12}$ in size. Then the series would become

$$1+ \tfrac{1}{2} + \tfrac{1}{3} + \tfrac{1}{4} + \tfrac{1}{6} + \tfrac{1}{12} = 2\tfrac{1}{3}$$

The only fractions allowed in the series are now ones that are multiples of $\frac{1}{12}$, like $\frac{1}{2}$ ($6 \times \frac{1}{12}$) or $\frac{1}{3}$ ($4 \times \frac{1}{12}$). We have tamed the infinite series and given it a finite sum. Similarly, by insisting that light energy should come in packets, rather than with continuously varying amounts, Planck was able to do away with the problem and restrict the light output to a finite total of energy. These packets, to which Einstein would give the name 'quanta', were not intended to be real. Planck didn't think that light came in chunks. That was silly – everyone knew that light was a wave, which should have continuously variable energy. What he thought he was doing was finding a patch that fixed the problem mathematically without necessarily reflecting the nature of reality.

As Planck later said, 'The whole procedure was an act of despair because a theoretical interpretation had to be found at any price, no matter how high that may be.' It's possible Planck should have thought a little more about history. The last time a major theory had used a model that was 'convenient for undertaking the calculations' but was thought to bear no resemblance to reality was when Copernicus suggested that the Earth travels around the Sun, rather than the other way around. And look how that turned out.

Planck was 42 in 1900, when he first proposed this theory. Not an old man, but already rather set in his ways. He would never be comfortable with the idea of light coming in chunks. But five years later, the then 26-year-old Albert Einstein was less worried about the preservation of old ideas. In a paper for which he later won a Nobel Prize, written in the same year as his famous paper on special relativity, Einstein took quanta of light at face value. He used them to explain the photoelectric effect.

When you shine light on to some substances, the light knocks electrons off the atoms, making an electrical current flow. If light were continuous, you would expect that the brighter the light, the more electrons you could knock out, whatever the light's frequency. Instead it was discovered that light below a certain cut-off frequency didn't knock out any electrons, however high you turned the brightness. This could be explained if light really did come in chunks. Unless you had a chunk with enough energy, it wouldn't matter how much light there was, there would be no dislodging an electron.

To make things even more uncomfortable for the likes of Max Planck, Einstein also showed that light could be accorded the same statistical approach as particles in a gas – something that fitted more comfortably with a collection of quanta (which would eventually in 1926 be called 'photons' by chemist Gilbert Lewis) than with a set of waves. That Planck was uncomfortable with Einstein's thinking was all too obvious when he recommended his younger colleague for membership of the Prussian Academy of Sciences in 1913. Planck requested that the academy wouldn't dismiss Einstein's application out of hand despite the fact that Einstein sometimes 'missed the target in his speculations, as for example, in his theory of light quanta.'

Into the quantum atom

Planck and Einstein's work was the beginning of something big – in fact arguably one of the two most fundamental parts of physics – and something that would show that the universe has randomness at its core: quantum theory. The next step down the road to randomness was taken by a man who, in some ways, was to Einstein what Einstein was to Planck – the Danish physicist Niels Bohr. Bohr turned Einstein's idea on its head and used it to explain the structure of the atom. Why, Bohr wondered, would an atom only absorb or give off light in chunks?

The idea that Bohr had was that the electrons buzzing around the outside of the atom travelled in orbits. This appealed strongly to a human tendency to build models based on observation. Bohr already knew that the planets orbited around the Sun – it would be very neat to have electrons orbiting around an atom in the same way. And it made for a simple, easily recognisable atomic symbol that is still widely used today, representing the atom as a miniature solar system. Such a pity, then, that this description of an atom was horribly wrong, something that Bohr realised almost straight away.

The trouble is that anything that changes direction is accelerating – and in an orbit, an electron would be constantly changing direction. Combine this with the fact that an accelerating electron is known to lose energy by pumping out light and you have an even bigger catastrophe than the ultraviolet one. If electrons really were like planets, orbiting atoms, the electrons of every atom would almost instantly lose their energy, causing them to spiral into the nucleus and that would be end of all matter.

Bohr fixed this by suggesting that the electrons around the outside of the atom could only travel in fixed orbits. It would probably have been better had he named his orbits 'tracks'. A satellite can travel in any orbit around the Earth, but Bohr's electron orbits were fixed at set distances from the nucleus. Rather than drifting in or out gradually, an electron had to jump instantly from one orbit to the next without passing through the space in between, making a quantum leap. If it jumped down it would give off a photon of light. To make it jump up, it would have to absorb a photon. This was a very satisfying model because it even explained why particular atoms only absorbed and gave off certain colours – because the energy of the photons corresponded to the size of the leap involved.

Bohr's ideas set off a revolution in physics with the likes of Prince Louis de Broglie, Werner Heisenberg, Erwin Schrödinger and Paul Dirac taking what was a relatively simple concept to help understand the structure of the atom and turning it into a whole new science of how tiny particles on the scale of photons, electrons and atoms operated. It was found that not only could light act like particles, but the components of matter could behave as if they were waves. There needed to be some way to pin down how these strange new particle/waves behaved and Schrödinger thought he had done this with an equation that was supposed to describe how something like an electron behaved over time. But once again, the theory produced an absurd result.

Schrödinger's wave equations predicted that since a particle like an electron also behaved like a wave, it should spread out over time. (Think of the kind of wave that is produced when you drop a stone into a pool of still water, rather than a wave on the sea. The wave is a circle, heading off in all directions. Schrödinger's equations reckoned particles should do this, but in all three dimensions at once.) Clearly photons and electrons and atoms didn't disperse like this – and yet the wave equations seemed to be such a good fit otherwise.

It was a friend of Einstein's, the physicist Max Born, who sorted out Schrödinger's problem. Where Schrödinger had assumed that his equations described where a particle was and what it did, Born thought that instead the equations told us the *probability* that a particle would be in a particular place. So instead of putting down an electron and having it embarrassingly ripple away in all directions, it meant that over time the electron could be in various different places, with the probability of the locations described by Schrödinger's equations.

It was a masterstroke on Born's part. It worked wonderfully, fitting observations with striking precision in all the experimental data that has since been gathered. But there was a massive price to pay. Born's quantum particle is not a distinct, specifically placed entity, but rather a fuzzy web of probability that gets more and more incoherent with time until a measurement is made.

The light revolution

This understanding of the nature of quantum particles was the key to cracking one of the biggest problems facing anyone trying to describe light as a collection of particles – Young's slits. This simple experiment had proved the death blow for Newton's idea that light was made up of particles and had confirmed the opposing theory of it being a wave. In 1801, Thomas Young had let a beam of light fall on two narrow, parallel slits. When the light continued through the back of the slits onto a screen, instead of forming two bright patches, as you might expect if light were made of particles, it formed a series of light and dark fringes.

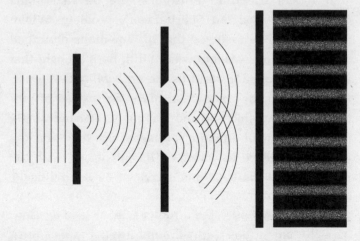

Young's slits

Young's explanation for this effect was that light was a wave, oscillating side to side as it travelled. When the beams from the two slits met, depending on the position they arrive at the screen, you could get both waves wiggling in the same direction – in which case they would reinforce each other and produce a bright fringe – or in opposite directions, in which case they would cancel each other out and leave the screen dark. There seemed no way to explain this effect if light was made up of particles. A beam of particles should produce two bright blobs on the screen, one behind each slit.

However, with Born's re-interpretation of Schrödinger's equation we can understand what is happening. Imagine sending single photons of light, one at a time, through Young's apparatus. This experiment has been performed many times since the idea first came up. Over time, as more and more photons pass through the slits, the familiar bright and dark fringes build up. This doesn't make sense if a photon is a little solid bullet of a thing. But if its position is described by a probability wave, it does not have a specific location. It is in a whole range of places at once, each with a different probability described by Schrödinger's equation. This means that it can pass through *both* slits, interact with itself (in a probabilistic fashion) and produce the required result.

If an experimenter decides to try to catch light out and puts a detector in one of the slits, the photons refuse to play ball. Even though the detector can be made to allow the light to continue through, the fringes disappear and we get the two bright patches you would expect from a series of conventional particles. The measurement made by the detector collapses the photon from being in a range of locations with varying probabilities, forcing it to go through a single slit and destroying the fringes.

A case of uncertainty

It is the same fuzziness of quantum particles that lies behind one of the most famous developments of quantum theory, Heisenberg's uncertainty principle. Like much of quantum physics, this has been painfully misused by those who like the words but can't be bothered to understand the underlying science, taking it to mean that nothing is certain, or anything can happen in the world. In fact, the uncertainty principle is firmly specific in its meaning. It says that quantum particles have pairs of properties that are linked. The more accurately you know one property, the less accurately you know the other.

There are several such pairs, but the best known is momentum and position. Momentum is the mass of the particle times its velocity. If we know a particle's momentum exactly, it could be located anywhere. If we know its position exactly, it could have any momentum. Usually, though, our knowledge sits somewhere in between, with a degree of knowledge and uncertainty for each property.

Although the uncertainty principle emerges from the maths, we can get a kind of sense of it by using the analogy of taking a photograph of a moving object. In taking the photograph I am, in a way, quantising reality, dividing it up into a series of 'snaps'. When I create one of these snaps I have a choice. I can take a very fast exposure. If I do so, the moving object will be frozen in time at an exact position. But I can deduce nothing about the way that the object is moving from the picture. It could be completely motionless for all I know.

Alternatively I can take a long exposure, say over 10 seconds. If the object is moving at any speed, then

I would expect it to appear as a long blur. If I measure the distance the object has travelled in that time on the photograph then I can work out how fast it is going. But in that snap I can't say exactly where it is – it is in all the places along the blur. This is only an analogy. In my blurred photo I could work out where the object was at a particular time, but I can't do this with a real quantum particle, where there is genuine uncertainty.

As more and more detail was added to quantum theory it was repeatedly backed up by experiment. Eventually the predictions of quantum electrodynamics (QED) would prove to be the most accurate match to experiment of any known theory. Richard Feynman likened the accuracy to knowing the distance between two cities on opposite sides of the United States to the width of a human hair. And yet quantum physics still confused and worried people. How could our solid, apparently dependable world be based on fuzzy probabilities? Was it really possible for everything we know to be based on particles that behaved in such a random fashion?

This dichotomy would be solidified in feline form by Erwin Schrödinger.

CHAPTER 11

No quantum cats

Even those most intimately connected with the nuts and bolts of quantum theory accept that it is nothing short of mind-boggling. Richard Feynman made this comment about his own specialist area of quantum theory, QED, in a lecture for the general public, emphasising that his audience should not worry that it's a struggle to understand quantum theory:

> What I am going to tell you about is what we teach our physics students in the third or fourth year of graduate school – and you think I'm going to explain it to you so you can understand it? No, you're not going to be able to understand it. Why, then, am I going to bother you with all this? Why are you going to sit here all this time, when you won't be able to understand what I am going to say? It is my task to convince you not to turn away because you don't understand it. You see, my physics students don't understand it either. This is because I don't understand it. Nobody does.

Scientists, Feynman suggested, sometimes just have to accept that as long as a theory successfully predicts what nature is going to do, it doesn't really matter whether that theory fits with our usual view of the world. He went on:

The theory of quantum electrodynamics describes Nature as absurd from the point of view of common sense. And it agrees fully with experiment. So I hope you can accept Nature as She is – absurd.

I'm going to have fun telling you about this absurdity, because I find it delightful. Please don't turn yourself off because you can't believe Nature is so strange. Just hear me out, and I hope you'll be as delighted as I am when we're through.

Not everyone, though, could go along with Feynman's easygoing approach. Some physicists agonised over quantum theory and what it meant. The sticking point seemed to be particularly the sharp distinction between the quantum world and the 'macro' world of everyday life. At the level of quantum particles, things could be in more than one place at once, could 'tunnel' through a gap without ever travelling across the space in between and could only be predicted in their behaviour by the random selections of probability – nothing was certain until it happened.

By contrast, in the macro world objects seemed not to be aware of this new understanding. Things still behaved the way they had before the new-fangled quantum theory came along. A ball, for instance, tends not to be in two places at the same time. It certainly can't tunnel through a solid object and its behaviour is predictable using Newton's laws of motion. And yet that ball is made up of quantum particles, all supposedly behaving in this weird fashion. How do we get a well-behaved ball from the probabilistic fuzz of the atoms that make it up?

In the quantum tunnel

We have already met the way that quantum particles can be in more than one place at a time, but let's examine the idea of tunnelling in a bit more detail too because it is a great example of how an apparently bizarre product of quantum theory is actually necessary to make the solid, predictable world work.

Newton was aware of one example of quantum mechanical tunnelling, though he didn't realise this was what was happening. If you shine light into a prism at the correct angle it will hit the back of the glass block and reflect back off, returning out of the front. This process is known as total internal reflection. It's straightforward, the sort of thing you do in school experiments, and usually no light comes out of the back. What Newton discovered (but they don't show you at school) is that if you put a second prism very close to the first but with a gap between them, then a weak beam of light emerges from the second prism.

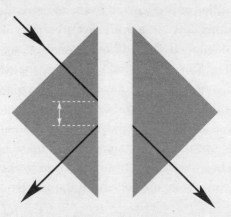

Frustrated total internal reflection

There was no possible explanation for this 'frustrated total internal reflection' in Newton's physics, but what is happening is that some of the photons of light are tunnelling across the barrier that is the gap between the prisms and carrying straight on in the second prism. Other examples of tunnelling have been discovered in a wide range of barriers. What has only been known relatively recently and shown experimentally is that this process involves zero tunnelling time – that the photon spends no time at all in the gap that it tunnels through.

The photon is, in fact, not travelling across the gap at all: it disappears from one side of the gap and reappears at the other. No time elapses in this process. (This means, incidentally, that a photon undergoing tunnelling makes its overall journey at faster than the speed of light, theoretically travelling backwards in time, but because tunnelling only works over very short distances, the resultant shift in time is far too small to be used in any way.)

If tunnelling only occurred in specialist equipment we might put this down as a minor weird aspect of quantum theory and ignore it. But without tunnelling we wouldn't be alive. The Sun is powered by nuclear fusion. For this to happen, hydrogen ions* have to be squeezed very close to each other. But the repulsion between their positive charges is so strong that even the heat and pressure of the Sun is not enough to get them to cosy up to each other enough. The only reason those hydrogen ions manage to fuse is because they tunnel across the barrier formed by

---

* A hydrogen ion is a hydrogen atom that has lost its single electron, which makes it simply a proton.

the electromagnetic repulsion. It is very unlikely to happen – but there are so many hydrogen ions that millions of tons undergo the reaction every second.

So this is a really strange aspect of quantum theory – particles that get through a barrier by disappearing on one side and appearing on the other without passing through the space in between – without which the Sun would not work and we would not be alive. It's hard to think of quantum theory as something abstract and irrelevant to the 'real' world when you realise this.

Dead or alive?

One of the founding fathers of quantum theory, Erwin Schrödinger, was so irritated by the difficulty of bringing together the probabilistic confusion presented by quantum theory and the solidity of the real world that he dreamed up what must be the most famous thought experiment in all of science – Schrödinger's cat. The concept has been described many times, but even if you are familiar with it, it is worth revisiting, just to emphasise what it is that has worried so many people.

We start with the observation that was made in the quantum version of Young's slits (see page 155). If you send photons through Young's slits one at a time you get the usual pattern of light and dark fringes. But if you put a detector in one slit, the photon is forced to go through one slit or the other and the pattern disappears. Generally speaking, a particle could be practically anywhere with the probability of it being in any particular place being given by Schrödinger's equation. But once you make a measurement and pin it down, the particle has to end up somewhere specific. The process was described as one of waveform collapse. The waveform, the outcome of Schrödinger's equation, collapses, to give a specific location.

The question that this idea provokes is just what making a measurement or 'looking at' the particle entails. One physicist, Eugene Wigner, suggested that it took a conscious mind to collapse the waveform, the consciousness somehow interacting with the particle's waveform. Although some still support this idea, many were doubtful from the start, and it prompted Schrödinger into action with his hypothetical cat.

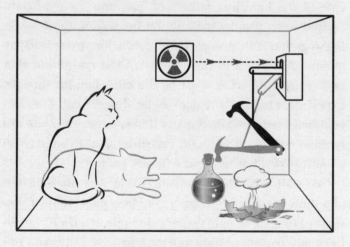

Schrödinger's cat

In the experiment (never carried out, it should be emphasised) a cat is placed in a box with a vial of deadly poisonous gas. There is also a radioactive particle in the box, which is due to decay at some point in time. When, we can't say. All it's possible to do with radioactivity is say that a particle has a certain probability of decaying, and with any particular population of particles, that half of them will decay within a known period (called the half-life). But we have no idea *which* particles will go in that period or when any particular particle will decay. We just know the probability.

Finally we have an ingenious little device, rather like a Geiger counter, that keeps an eye on the particle and notices when it decays, picking up the radiation produced. It then triggers a Heath Robinson device that smashes the vial, releasing the deadly gas and killing the poor cat.

Here's the thing. We close the box and wait a bit. The particle is now in two states at once. Just as Schrödinger's equation tells us that over time the probability of a particle being at a distance away from its original position increases, similarly it tells us that over time, the probability of the radioactive particle having decayed increases. With some time elapsed, the particle is both in the undecayed state with one probability and the decayed state with another. Because it's a quantum particle, there is no hidden value – it isn't in one state or the other. The actual state doesn't come into existence until the particle is observed. If the particle is both decayed and not decayed, this seems to suggest that the detector is both triggered and not triggered. And so the cat is both dead and alive. This is worrying.

If quantum theory were left to its own devices, the moment the box is opened to take a look at the cat, the waveform collapses and either one or the other state is revealed. But until that point in time we had a dead-and-alive cat. Many would argue that the presence of the detector is enough to collapse the waveform and decide the fate of the cat. But those following Wigner's lead would genuinely believe that the cat was in both states.

More recently, cats around the scientific world have been able to heave a collective sigh of relief thanks to a concept called decoherence. This essentially says that when a quantum particle comes into contact with the environment around it – typically by interacting with another quantum particle – the result is the appearance of the collapse of the waveform. There are plenty of quantum particles in the detector and the box, and their mere presence is enough to result in decoherence and a shift from pure probabilities to reality.

There is a subtle technical difference between decoherence and a true waveform collapse – what happens in decoherence is that the particle's waveform becomes effectively tangled up with those of the particles in the environment around it. But the result is that the particle behaves in a more predictable macro manner as if the waveform has been collapsed. It loses much of its quantum weirdness. Because of this, decoherence is something that those trying to build quantum computers (see page 185) are always fighting, as they need to keep their quantum particles in a pure probabilistic state.

Interpreting the quantum

In some way or other, the split between the weird quantum world and the boring, predictable macro world has to be artificial. Otherwise, where would you draw the line? We know one atom acts in a quantum fashion, as do two. Physicists have managed to perform the Young's slits-style 'in two places at once' experiments with buckyballs. More properly known as buckminsterfullerene molecules, these are near-spherical carbon molecules with 60 atoms arranged in a structure that looks unnervingly like a football. It has been suggested that it would be possible to achieve the same effect with a small virus. Where does the break come between this and normal 'macro' things?

This is rather different from the difficulty facing earlier scientists trying to explain the nature of gases. They could get away with using a statistical approach, simply reflecting the unmanageable scale of the calculations and data required otherwise. There was no way that you could measure and follow every atom in a classical, pre-quantum gas. Not only would it take an impossibly long time to work out, it wasn't practical to do. But the statistical approach let you take the overview, getting a feel for what was happening on the macro scale, aware that your numbers were based on a solid foundation of reality below at the level of atoms. With the advent of quantum theory, that foundation was swept away and you were left with a castle in the air.

The original approach to dealing with this, still accepted by many physicists, is simply to shrug the shoulders and say, 'Oh, well, that's the way it is. The numbers work – as long as they do, we can ignore what's underneath and accept that we don't (and never can) understand it.' This attitude is at the heart of the

Copenhagen interpretation, developed in 1927 by Niels Bohr and Werner Heisenberg to try to paste over the cracks of quantum theory. The Copenhagen interpretation simply says that there is nothing meaningful until the measurement is taken. There is no point worrying about where a particle is or what it is doing, it's just a probabilistic mess. Nothing really exists until the measurement is made.

There are a whole collection of other possible explanations and interpretations, many of which are variants of Copenhagen. Personally I am entirely comfortable with the implications of Copenhagen, even if you have to allow for added concepts like decoherence. I quite like the idea of just letting go of what is 'actually' happening beneath the surface at the quantum level as irrelevant and something we don't have to worry about. I even quite like the idea of that randomness bubbling away under the surface. But not everyone is comfortable with this. As we will see in the next chapter, Einstein was horrified by it. And some physicists think it is more likely that a very different interpretation of quantum theory explains what is happening better: the only problem is, for it to work, you have to accept there is not just one reality but many, many parallel versions of existence. This interpretation is usually called 'Many Worlds'.

Many Worlds dates back to 1957 when a young scientist called Hugh Everett decided that he couldn't accept the Copenhagen view and wrote a controversial PhD thesis to show how it could be evaded. In Many Worlds, each time there is a quantum 'decision' the universe splits into two. Both versions of the universe then exist in parallel. As this happens as a result of every inter-action of every quantum particle in the universe, there are indeed many worlds in this picture.

In some ways, Many Worlds is attractive. We no longer have to worry about how a photon can be in two places at once and go through both slits in Young's experiment. Instead it is reality that splits. In one universe the photon goes through the left-hand slit and in the other it goes through the right-hand slit. Although we can only ever experience one universe – which is why a detector pins the photon down to a single slit – we see the result of these different universes interacting in the interference pattern of light and dark fringes that is produced.

For the Many Worlds enthusiast, Schrödinger's cat holds no mystery. In one world the cat is alive, in the other it is dead. We are in one branch of the universe and so experience one state for the cat. It is never alive and dead at the same time in a single universe. In the Many Worlds version of reality, waveforms don't col-lapse because all possible outcomes exist, each in its own branch of reality.

For some observers, Many Worlds adds a ridiculous level of complexity simply to deal with the discomfort raised by an interpretation. Common sense suggests that a universe that multiplies with such rabbit-like fecundity is unlikely (though common sense rarely gets things right in quantum physics). Each interpretation has advantages, some mathematical, some in spirit. Many Worlds would be useful, for instance, if backward time travel were pos-sible to get around the paradoxes that arise. But as yet there is no evidence from experiment to favour Many

Worlds – in fact no way to determine which interpretation is closer to reality, other than simple preference.

You will read in some accounts that Copenhagen has fallen out of favour and that the majority of physicists prefer Many Worlds. I just don't believe this. There is a core of Many Worlds enthusiasts who are very vocal, but my suspicion is that outside that cadre, the vast majority think that Many Worlds is interesting but can't say they wholeheartedly support the interpretation. I am with them.

Whatever the interpretation, the scary thing about quantum theory is that it makes randomness a reality and bases existence on the action of probability. With our obsession with patterns and expectation of solid everyday behaviour, this sits uncomfortably with many people. And none more so than one of the people who laid the very foundation of quantum theory, Albert Einstein.

CHAPTER 12

Improbable world redux

Think of the most revolutionary scientist of the 20th century. You don't have to stretch your imagination too far. The chances are that you settled on Albert Einstein. Why wouldn't you? Einstein didn't just sort out the photoelectric effect by assuming that quanta of light were real. He also explained Brownian motion, producing the first proper understanding of the reality of atoms.* He came up with special relativity, showing how Newtonian physics has to be modified when things are moving quickly. And he developed general relativity, finally explaining why gravity works – a far cry from Newton's basic mathematics to predict how strong the pull of gravity will be.

We even have those familiar images of Einstein – the wild-haired old guy with his tongue sticking out – that make it clear he was in every sense an iconoclast. So it's remarkable to have to admit that this revolutionary, highly original thinker simply couldn't accept that quantum theory – and hence reality – had probability at its heart. Just as Max Planck was uncomfortable with the consequences of his theoretical quantum of light turning

---

* Brownian motion is the jerking around of small particles such as pollen grains suspended in a fluid. It is caused by the jiggling atoms or molecules in the fluid bashing into the pollen grains. It's hard to believe now, but when Einstein wrote his paper on Brownian motion in 1905, many scientists did not believe that atoms really existed, thinking of them as useful models rather than a true facet of nature.

out to be real, Einstein was horrified that his real quanta dragged probability into the natural world. For many years, even while he was busy overthrowing many existing ideas about gravity with general relativity, Einstein would spend his spare moments searching for the error that would bring quantum theory crashing down.

You only have to take a look at some of his letters to his friend and fellow physicist Max Born. In 1924 he wrote:

I find the idea quite intolerable that an electron exposed to radiation should choose of its own free will, not only its moment to jump off, but also its direction. In that case, I would rather be a cobbler, or even an employee in a gaming house, than a physicist.

A couple of years later he added:

Quantum mechanics is certainly imposing. But an inner voice tells me that it is not yet the real thing. The theory says a lot, but does not really bring us any closer to the secret of the 'old one'. I, at any rate, am convinced that He is not playing at dice.

This is one of the originals of the quote that's usually given in the condensed form 'God doesn't play dice.' What upset Einstein was that it seemed as if there was no underlying cause behind the decay of an atom or other probability-driven quantum events. There was nothing to instruct a radioactive atom when to decay, or a photon which of Young's slits to go through. For Einstein this was intolerable. He was convinced that there had to be something hidden in nature that we just couldn't access, information that would enable us to know when and how a quantum event would happen. It was hidden, perhaps irretrievably so, but still there.

Take a classic problem that has puzzled physicists since Newton's time. It's a fantastical quantum mechanical event that you can test out for yourself and that illustrates very clearly why the whole business got Einstein so irritated. All you need for your experiment is a room with an electric light, a window and night-time.

The quantum mechanical window

Stand in a room with the lights on at night and look at the window. You will see a reflection of yourself and the room. The darkness is making the window glass act like a mirror. Now go outside and look at that same window again. You can clearly see into the room. Let's get a feeling for what's happening here. Some of the light from the room is passing through the window. It must be, or you wouldn't be able to see in from outside. Some (a much smaller part) is reflecting back into the room, so you can see yourself reflected inside. This happens all the time – the darkness doesn't make it happen, it just means that the reflections aren't drowned out by light that pours in from outside during the day. It's the same as seeing stars. The stars are 'out' in the sky all day, we just can't see them because they are drowned out by sunlight.

Now let's imagine what's happening at the level of individual photons of light. A photon hits the surface of the glass. It might pass straight through, or it might be absorbed and re-emitted back the way it came (what we usually think of as reflected). Some of the photons hitting the surface – let's say 10 per cent of them – will go back into the room, some – in this case the remaining 90 per cent – will pass through. Which is fair enough. But how does a particular photon know what to do, whether to reflect or to carry on? We can say with confidence that 10 per cent (or whatever the number really is) of the photons will reflect. But any particular photon must do one thing or the other. There is nothing about a photon that tells you how it will behave. The decision to reflect or not to reflect is purely probabilistic.

Newton had a problem with this, as he correctly believed that light was made up of particles, which he called corpuscles. He tried hard to come up with a reason why any particular particle would decide to reflect or pass through the glass. The only thing he could think of was that there were imperfections in the surface of the glass – that effectively it was the minute scratches, bumps and dirty bits that caused the reflections. In Newton's model, a perfectly smooth, untarnished piece of glass would not reflect any photons back into the room. Unfortunately for Newton, this has proved not to be the case. It doesn't matter how perfect your glass surface is, you will still get reflections.

However, this isn't the weirdest aspect of the whole business of a beam splitter, which is the technical term for what a sheet of glass does in these circumstances. If you use different thicknesses of glass you will find that a higher or lower percentage of the photons will reflect from the inner surface of the glass, the one inside the room. (There are also reflections from the back surface, where the glass meets the outside air, but we'll ignore these.) So somehow, not only does a photon hitting the inner surface have to decide whether or not to reflect, it also has to know in advance how thick the glass is, to get the probabilities correct. Weird indeed.

To understand just what is going on in that piece of glass (who'd have thought a simple window could be such a sophisticated piece of technology?) we have to return to the masterpiece of one of the greatest physicists who ever lived, Richard Feynman. We have already seen (page 40) his idea of using the sum of all possible paths a particle could take. This was part of his approach to Quantum Electrodynamics (QED for short), the science of the interaction of light and matter.

Generally speaking, a photon of light is created when an electron in an atom drops in energy (a downward quantum leap). The photon carries away the energy that the electron has lost at the speed of light until it interacts with another electron. This could be in an adjacent atom or after the photon has crossed millions of light years of space. The electron absorbs the photon, jumping up a to a higher energy level. This is the basis for most inter-actions between light and matter.

When a photon is absorbed by an electron in the win-dow glass it could then be re-emitted in any direction – in fact taking Feynman's approach to QED it actually *is* emitted in every possible direction, each with a different phase and a different probability. When we then add up all the different paths many of them cancel out and we will typically end up with a simple reflection just as if light had bounced off a mirror – though in reality, photons aren't doing any bouncing at all.

What happens in our window glass is that we have to add together the possible interactions and possible paths from a photon interacting with electrons all the way through the glass. This is why the thickness has an effect on the reflection from the inner surface. Differing thicknesses will provide differing phases of re-emitted photons to cancel out other possibilities. So it is the potential interaction of photons and electrons all the way through the glass that decides on the final outcome of 'reflection' from the inner surface. And always we have to remember that we have no way of predicting what any specific photon will do in each potential interaction. Each is just a probability, adding together to produce the final outcome.

Einstein's hidden truths

Einstein was sure there had to be a more reasonable explanation for what was going on in nature than the assertion 'Randomness is in charge'. It's not that Einstein had any problem with probability and statistics per se. He had made plenty of use of statistics in his work on atoms and on light treated as a gas. But for him, the probabilities that emerged should reflect underlying truths, rather than being a fundamental part of nature.

Take a very simple example. Imagine we've a room containing 100 men and 100 women, all milling about. You give them a chance to get well and truly mixed up, then open a door and pick the nearest person out. There is a 50 per cent probability that the person you pick will be a woman. But that probability doesn't suggest any-thing inherent about the person. She wasn't 50 per cent male and 50 per cent female until you made the observa-tion. There was always an underlying fact that the person was a woman, though it was hidden from you until you opened the door. This is how Einstein believed quantum particles must be.

For a number of years, Einstein threw a whole range of thought experiments in front of the great prophet of the quantum, Niels Bohr. Each was intended to bring the nature of quantum theory into question. Initially Bohr was baffled by the whole business, but over time it seems to have developed into a pleasantly anticipated game between the two of them. They would meet at some conference – quite possibly nothing to do with quantum theory. Over breakfast, or in some other casual meeting spot, Einstein would present Bohr with a challenge that he thought demonstrated the flawed nature of quantum physics. Bohr would ponder the thought experiment for a few hours, then show Einstein where he had gone wrong.

Einstein's final casual challenge in this series came in 1930. But for the following five years, a spare time activity for the great man was putting together his ulti-mate challenge to quantum theory's apparent foundation on randomness. Like most of Einstein's challenges it was in the form of a thought experiment, a test that did not need to be carried out, but that could be used to probe the theory. And this time Einstein thought he was onto a winner. Rather than pose his experiment over break-fast, he published it formally as a scientific paper. It was titled *Can Quantum-Mechanical Description of Physical*

Reality Be Considered Complete?, but thankfully is usually referred to by the initials of its authors, Einstein, Podolsky and Rosen – EPR.

The premise is relatively simple. We produce two quantum particles in a way that intimately ties them together. There are a number of ways of producing this state, known as entanglement – the simplest is if an electron dropping down an energy level in an atom produces not one but two photons, which will be entangled. A more controlled approach is to make use of beam splitters in a rather more sophisticated arrangement than a piece of window glass, which can then be used to create entanglement.

These two entangled particles shoot off in opposite directions. We allow them a good amount of time to get well separated, then make a measurement on one of the particles. The original paper used the momentum and position as properties to be measured, which confused many into thinking this was a challenge to Heisenberg's uncertainty principle. This was never the intention. In fact Einstein, commenting on the choice of momentum and position, said 'Ist mir Wurst' (literally, 'It's sausage to me', roughly equivalent in colloquial German to 'I don't give a damn'). The idea was just to measure a single property that entanglement meant was intimately linked between the particles. By demonstrating the effect with two properties rather than one, the original paper caused a lot of confusion.

More modern developments of the EPR experiment usually choose instead the property known as quantum spin. Quantum particles have this property, which isn't really about spinning around – the name just stuck because it was initially thought this was the case. Quantum spin is a property that can only have one of two values in any particular direction. If you measure the spin in the vertical direction, say, it can only either be 'up' or 'down'. This spin is a bit like the photon's chances of reflecting off window glass. Before measuring it, we don't have any idea what the spin will be, we can only know probabilities. It might be that we know that there is a 50 per cent chance of the spin being up and a 50 per cent chance of it being down. It is only when the measurement is made that the particle gets the particular value – up, say. Before that there was nothing hidden away to say what the outcome would be.

Now here's the interesting bit. Instantly, as soon as the first particle is measured to be spin up, then the other particle that is entangled with it will become spin down. It too was in both states until that point – now it takes on a specific value. And this happens instantly at any distance. The second particle has to become spin down to conserve spin, which can't be created from nowhere any more than energy can – but no one knows how the information gets from one particle to the other.

This seemed to contradict Einstein's special relativity, which made it clear that information shouldn't be able to travel faster than the speed of light. So, Einstein declared in the paper, there were two possibilities. Either quantum theory was wrong and there was hidden information that determined the state particles would be in ahead of the measurement, or you had to do away with locality, the idea that a particle was in a particular location and could not communicate with another, distant location faster than the speed of light. The paper's response to the possibility that locality could be got around was 'No reasonable definition of reality could be expected to permit this.'

It took Bohr weeks to respond to this paper, and when he did he seems to have been confused by the paper's muddled wording. In a way, he couldn't challenge the conclusion that either quantum theory was wrong or it was possible to breach the local nature of reality – but Einstein's dismissal of the challenge to locality was arbitrary and not based on any science.

There is only so far you can go with a thought experiment when it comes to testing such an 'either/or' conclusion. By the 1960s, though, the British physicist John Bell had come up with a measurement that would make it possible to test between the two possibilities, and two decades later, French scientist Alain Aspect had put together an experiment to carry out that test. Bell's approach depended on a subtle difference that would be observed in photons depending on whether there was hidden information or an instant communication, but the outcome was clear. Aspect demonstrated that entanglement really did break the local reality barrier and could communicate instantly across a distance. Any distance.

Since then there have been many experiments making use of entanglement and they have consistently demonstrated that Einstein was wrong on this matter. He didn't live to see it happen, but the experiments confirmed that there were no hidden values tucked away out of sight to steer reality. Quantum particles really do exist in a probabilistic state, only deciding which way to go when they interact with something else, whether it's measurement of a property or a simple reflection off window glass. At the level of quantum particles, randomness truly does rule.

Quantum secrecy

Entanglement would be fascinating even if it only allowed us to demonstrate the role of randomness and probability in quantum reality, but it has much more to offer. I won't go into too much detail (there's a lot more in my book *The God Effect*), but just to take three remarkable possibilities it offers, entanglement gives the opportunity to provide unbreakable encryption, provides an essential mechanism for building quantum computers, and forms the basis for miniature *Star Trek*-style transporters.

Encryption is the business of creating codes and ciphers to keep information secret. It may seem like the stuff of spies and the military – and it's true that secrecy is an essential in those fields – but they are relatively small-scale users of the modern technology of secrecy. These days it's much more likely to be a feature of banking and commerce. When you log into your bank via the web, or purchase something from an online store, somewhere tucked away at the top or bottom of the browser there should be a little padlock symbol, indicating that this is a secure transaction.

What security means in this context is that the data that travels between you and the bank or shop is encrypted, is sent in a coded format. It's very easy to make an unbreakable code*. They have been around for

---

* Technically what I'm talking about here is a cipher, not a code. A code substitutes a special word for another word or phrase. So

about 100 years. All you need is what is called a one time pad. This provides you with a series of random values (yes, it's another application of randomness), which are added to your actual message.

Imagine, for example that I wanted to send the message

DICE WORLD

Then my one time pad gives me a series of random numbers to add to the letters. (If letter arithmetic seems a little strange, all I do is move further along the alphabet by the number specified. When I get to the end of the alphabet, I carry on at the start, so A is the next letter after Z.) Let's give that a try. I am using my trusty Excel pseudo-random number generator (see page 128) to come up with a string of numbers, one for each character. My numbers are:

8, 13, 16, 22, 12, 13, 26, 2, 3, 17

To be particularly sneaky, I am treating a space as a 27th character, so I will lose the word distribution in the original to make the message more cryptic. Add those numbers to the text and I get:

LVS LINTOU

I might have a code book that tells me that the word 'antioxidant' in a coded message meant 'Send reinforcements, we're going to advance.' A cipher is a character-by-character substitution of different values in a message like the one described here. But in general English, 'code' is often used for any encryption, while a cipher tends to be a mystery that turns out to be of no importance.

The clever thing about this is that these are now random characters (at least, as random as my random number generator allows them to be). So there is no way to break the cipher by looking for any kind of structure – there isn't any. But there is also a huge drawback with one time pads. The only way to decrypt the cipher is if the person reading the message also has a copy of the one time pad values to subtract from the letters and get back to the original. And as soon as I send that key from one place to another, whether as a literal paper pad or electronically, I risk it being intercepted and my message being read.

This was the reason that the famous Enigma machines used by the German military in the Second World War didn't use an unbreakable cipher, just one that was very hard to break. Relying on a mechanism rather than a one time pad meant the machines could be used at two locations without a key being sent from place to place. The same goes for your internet banking. It would be very inconvenient if you had to have a special set of information provided to you in a secure way every time you wanted to login, so instead, the software makes use of a means of encryption that is very difficult to break, but that can be used remotely without passing a known key from place to place*.

---

* The security mechanism often used on computers does involve passing a key around, but this is a rather special form of encryption where you use one key, the so-called 'public key' to encrypt the information and another key, the 'private key' to decrypt it. This asymmetric approach is clever because you can give your public key out freely, but no one except you can read a message that is being sent to you, as that requires the private key, which you keep to yourself.

Entanglement, though, offers the possibility of getting all the unbreakable security of a one time pad approach without the possibility of the key being copied. Say our sender used particles that had a 50:50 chance of having a spin up or a spin down. She produces a set of such entangled particles, keeping one of each pair at the sending end of the communication chain and one at the receiving end. The sender now starts to examine the particles. She uses spin up as 0 and spin down as 1 to provide a random key which is added to the message. (The message itself would be in binary, so the key only needs to have two possible values). The message is then transmitted by conventional means – radio, say – as a totally random set of characters.

The receiver then simply uses the values of his own particles, but in reverse. For him, spin up is 1 and spin down is 0. He subtracts the values and the message is decoded. The great benefit here is that the one time pad does not exist until the message is sent. It is generated by the sender at the moment of encoding – and only comes into existence for the receiver the moment that the message is being prepared. No one could sneak a look at the pad in advance and then use it to decode the message, because it doesn't exist ahead of time.

The one possibility to get around this mechanism is that someone could intercept the entangled particles that are sent to the receiver before the message, triggering the establishment of the key ahead of time. That way it might seem that someone could get hold of the key. But there is a mechanism to check if particles are still entangled. It requires extra information to be passed between sender and receiver, but it is entirely possible. If someone intercepted the key, they would break the entanglement – and so by checking every few particles to see if the entanglement is still intact, the link could be taken down as soon as there was any chance of the message being intercepted.

Computing with quanta

Ironically, the second application of entanglement is a kind of mirror image of this quantum encryption, as it is potentially a way to *break* our current means of keeping data secure. This would be possible if we had a working quantum computer. In the kind of computer we use every day, information is stored as bits. These can have one of two values, usually represented as 0 or 1. Every action the computer takes involves finding bits, moving them around or flipping them. A quantum computer replaces bits with qubits – quantum bits – each qubit of information being the state of a quantum particle.

We have already seen how a quantum particle can be in two states at once – a superposition. Think of a particle's spin before we measure it. When the measurement is taken in a particular direction, there will be a probability of it coming out up and a probability of it coming out down. Before making the measurement, we can only say that it is, say, 43.5 per cent up and 56.5 per cent down. This is effectively the value of the qubit corresponding to spin. And because the qubit can take any value between 0 per cent up and 100 per cent up, in effect this is a 'bit' that can hold a real number. Not just 0 or 1 but any value between 0 and 1. With the values I picked, that would make the number 0.435 (or its complement 0.565 – either could be chosen).

By using these values in calculations you can perform much quicker processing than with a conventional computer, because you aren't handling individual bits at a time. What's more, it can be much more accurate. As we saw on page 92, chaotic behaviour was discovered when a real number was rounded from the six digits that were used by the computer to the three digits on the printout. But in a quantum computer, in theory a number can be held to an infinite set of decimal places. However, quantum states are fragile. To keep a qubit in a particular state means isolating it as much as possible with contact with other particles to avoid decoherence. This is possible, but not necessarily for very long. One way to handle this is to treat qubits like hot potatoes, passing the value from qubit to qubit before it collapses.

Even if you can keep the values stable, it's extremely fiddly getting information in, out and around such a computer. Any attempt to examine the value of a qubit will change it. And this is where entanglement comes in. It seems pretty well impossible to construct a quantum computer without making use of entanglement to pass information around. Entanglement provides a means to make use of a qubit's properties without ever actually finding out what those values are and corrupting them.

Many teams around the world are working on different designs of quantum computer. Progress is slow, but there have been models put together with a handful of qubits which can perform basic arithmetic. However, if a reasonable scale quantum computer can be put together it would be capable of remarkable things. We know this because the basic structure of at least two powerful algorithms that would only run on quantum computers – effectively mathematical recipes – have already been put together.

One makes it possible to find out which two prime numbers* have been multiplied together to produce a third, very large number. If the resultant number is big enough it is actually very difficult, even for the fastest current computers, to work out from the result which prime numbers were multiplied together to produce that value. This might seem like the kind of thing that could only excite a mathematician (they are easily excited) – but it is of huge significance because of the RSA algorithm.

RSA is the mechanism used to keep information like your credit card details secure when transmitted over the internet. It is RSA that makes use of the public key/private key approach described above. The RSA mechanism relies on distributing the result of multiplying two very large prime numbers together – the public key – but keeping the knowledge of those two primes secret. In principle RSA is breakable – but if you make those primes big enough, the calculation to work them out would take years to run with the fastest supercomputer. By comparison, a quantum computer could make that calculation in seconds. If we ever get fully functional, commercially available quantum computers there will

---

* Prime numbers, as you no doubt remember from school, are numbers which can only be divided by themselves or 1. The first few are 2, 3, 5, 7, 11, 13, 17 ... (It is arbitrarily decided that 1 itself isn't a prime number.)

have to be a serious rethink of the way information is kept secure on the internet.

A more positive application of quantum computers, which would have Google and other search engine suppliers rubbing their hands with glee if it became available, is the Grover Search Algorithm, which has been described as the 'needle in a haystack' program. This particular computational recipe cuts through what can be a painful search problem when you are dealing with vast quantities of information. A typical application would be a directory. It's easy to look someone up by name in an old-fashioned paper phone book. But it's a real pain to use that directory to find a person, given a number.

At the moment, computers have two approaches to this. One is the index. This is a bit like looking up someone by name – you have a list of names, and for each you have a pointer to the full information. You could also set up a second index organised by phone number, and that way you could quickly find a person given the number. But this only works if you have already set up an index, a time-consuming process. If you have lots of messy information – websites, say, or documents – it's a vast amount of work to index every single word. Often a computer will have to rely on brute force. In the example of the phone book, it would have to look at every entry until it finds the right one.

In practice, with the relatively small amount of data in a phone book (by computing standards) a computer could do the job quickly – but if you were trying to access world-wide-web-sized quantities of data, it would be a different matter. Keeping the example of the phone book for simplicity, let's imagine it has a million names and numbers in it. If there is no index, using the brute force method, the computer will have check up to 999,999 values before it finds the correct one. On average, it would take 500,000 checks before hitting the number that was required. But the Grover Search Algorithm means that a quantum computer could find the right entry with just 1,000 searches – the square root of the number of entries. When you take such capabilities up to the many billions of chunks of data on the net it becomes a very attractive algorithm indeed.

Beam me up

The final application of entanglement worth a mention here is quantum teleportation, which provides a small-scale version of the *Star Trek* transporter. Think for a moment of what's involved there. On *Star Trek*, the transporter appears to scan an object or person, then transfers them to a different location. This was done on the TV show to avoid the cost of the expensive model work that was required at the time, before the existence of CGI, to show a shuttle landing on a planet's surface. But it bears a striking resemblance to something that is possible using entanglement.

To make a transporter work, we would have to scan every particle in a person, then to recreate those particles at a different location. There are two levels of problem with this. One is an engineering problem. There are huge numbers of atoms in a human body – around 7×10^{27} (where 10^{27} is 1 with 27 zeros following it). Imagine you could process a trillion atoms a second. That's pretty nippy. But it would still take you 7×10^{15} seconds to scan a whole person. Or to put it another way, around 2×10^8 years. 200 million years to scan a single person. Enough to try the patience even of Mr Spock.

Assuming, though, we could get over that hurdle – or only wanted to transport something very small like a virus – there is a more fundamental barrier. What you need to do to make a perfect copy of something is to discover the exact state of each particle in it. But when you examine a quantum particle, the very act of making a measurement changes it. It isn't possible to simply measure up the properties of a particle and make a copy. But quantum teleportation gives us a get-out clause.

There is a slightly fiddly process using entanglement that means we can take the quantum state of one particle and apply it to another particle at a different location. The second particle becomes exactly what the first particle had been in terms of its quantum properties. But we never discover what the values are. The entanglement transfers them without us ever making a measurement. Entanglement makes the impossible possible. This has been demonstrated many times in experiments that range from simple measurements in a laboratory to a demonstration that used entanglement to carry encrypted data across the city of Vienna.

Even if we were able to get around the scanning scale problem, though, it's hard to imagine many people would decide to abandon cars or planes and use a quantum teleporter for commuting. Bear in mind exactly what is happening here. The scanner will transfer the exact quantum state of each particle in your body to other particles at the receiving station. The result will be an absolutely perfect copy of your body. It will be physically indistinguishable, down to the chemical and electrical states of every atom inside it. It will have your memories and will be thinking the same thoughts. But in the process of stripping those quantum properties, every atom of your body will have been scrambled. You will be entirely destroyed in the process. As far as the world is concerned you will still exist at the remote location – but the original 'you' will be disintegrated.

This is not ideal if you are looking for a worry-free way to travel, but quantum teleportation does offer all sorts of possibilities for communicating information in different ways and also provides valuable assistance in getting information around a quantum computer.

All these applications are real and have been widely tested. They blow apart Einstein's concern that there is no dice world. If there were truly the hidden reality that Einstein was sure had to lie beneath quantum randomness and probability, then entanglement could not perform the remarkable feats it does.

When Einstein and his colleagues came up with the EPR paper, predicting that quantum theory implied the possibility of entanglement, they intended it as a death blow for quantum physics. The 'spooky' link of entanglement, as Einstein referred to it, would be thrown away and with it the idea that reality, solid everyday reality, was based only on probabilities. But Einstein, Podolsky and Rosen failed in a resounding fashion. Einstein may never have liked it, never accepted it, but the randomness that lies beneath everyday existence was here to stay.

If we contrast the modern physics that emerged in the 20th century as a result of relativity and quantum theory with what came before, it's like the difference between spaceships and steam engines. And yet the fundamental theory behind the way steam engines worked was itself a demonstration of the lack of certainty in the physical world. If Einstein was worried by the behaviour of the quantum, his predecessors were challenged by issues that emerged with the hissing exhaust of the steam engine. The laws of thermodynamics.

CHAPTER 13

Follow the heat

While quantum theory was changing our view of matter and light, another apparently staid physical theory from the Victorian era was beginning to boggle the minds of observers – and once again it was the presence of probability and statistics at the heart of reality that caused the confusion. The culprit was thermodynamics. Thermodynamics began as a basic understanding of the working of steam engines, but one apparently simple rule – that when you put a hot thing and a cold thing in contact, the hot thing will get colder and the cold thing hotter – proved to be a statistical can of worms.

Before we dig deep into that pulsating container we need to understand a little about what thermodynamics is, and along the way it is important to shrug off a Victorian hangover in language. Because everyone else does, I'm going to refer to the 'laws of thermodynamics', but that 'laws' word makes me cringe. That's because the concept of a 'law of nature' is painfully unscientific and out of date.

Science is a mechanism for understanding how the universe around us works, and on the whole, science works superbly well. But it is easy for science, and those talking about it, to get an overinflated sense of its own importance. In very broad terms, science works by putting up a theory, testing that theory against the available data and if necessary disposing of the theory or modifying it if it doesn't match reality.

Our scientific theories provide our current best guess at what's going on. But they are only as good as the next bit of data. So, for, instance, Newton's 'laws' worked fine with the basic data available at the time, but had some problems with special cases – when, for example, something is moving very quickly or is in a powerful gravitational field. So Newton was modified to come up with special and general relativity – Einstein's improvement on Newton. And at the moment, relativity provides our best guess. But that too could well need modifying, or even scrapping entirely, in the future, to deal with new data.

For me that term 'law' smacks far too much of Victorian certainty. Science can never tell us definitively the facts of how things are; it can just provide our current best theories given the data. That's a whole lot better than anything else can do. It's not a licence for saying that anything goes. Because science does give us the *best* guess. But science has this limitation, which even scientists often forget about. So, for instance, you will hear a cosmologist speaking as if the big bang is fact. It isn't. It's our current best theory. And that's all you are ever going to get.

In reality, then, the 'laws of thermodynamics' should be the 'theories of thermodynamics.' But we are stuck with the term and might as well get on with it. There are four laws of thermodynamics in all. We will be particularly interested in the second law – but it's worth getting them all in place to put the whole set into context.

Laying down the law

We start with the zeroth law, so called because it was tacked on after the first and second law were already established, somewhat like my zeroth chapter. The zeroth law says that two objects will be in equilibrium (balanced) as far as heat is concerned, if heat *can* flow from one to the other, but it doesn't. If two objects, each at the same temperature, are in contact with each other, one won't influence the temperature of the other. In practice there will be a constant flow of energy backwards and forwards between the two objects, but what the zeroth law really means is that the net flow is zero. Another way of putting it is that if A is in equilibrium with B and C is in equilibrium with B, then A and C are also in equilibrium.

The first law stems from the conservation of energy. It just says that the energy in a system changes to match the work it does on the outside (or that's done on it), and the heat given out or absorbed. These are all forms of energy, so basically it tells us that the total energy in a closed system will stay constant*. You can't make energy or lose it. In the real world, we don't make energy with power stations and generators and the like, we harvest energy that is already there.

The second law of thermodynamics is also about heat going from place to place. It says that, when left to its own devices, heat moves from a hotter part of a system to a cooler part. That sounds simple and commonsense, but it has profound consequences, all the way up to the

---

* With Einstein in mind, we really mean that the total mass and energy combined in a system will stay constant.

future of the universe. As we will discover, another way of stating the second law is that, in a closed system, entropy (see below) will stay the same or rise. The third law has little impact on our everyday lives. It says you can't get a body down to absolute zero (0K, −273.15°C or −459.67°F) in a finite number of steps. You can always get a fraction closer, whatever temperature you are at, but you can never quite make it. Getting to absolute zero is like adding ½ + ¼ + ⅛ ... and so on. In the real world, that series will never quite make a total of 1.

The word 'system' has cropped up a lot in these quick summaries, and it is an absolutely fundamental concept for understanding thermodynamics. The system is the bit of the universe we are looking at for thermodynamic purposes. For early thinkers about thermodynamics, who were interested in making their technology more efficient, the system would typically be a steam engine – but it can be absolutely anything. A cup of tea. You. Your house. The Earth. The entire universe. Systems in the real world (as opposed to the imaginations of theoretical physicists) are mostly open – which means they can interact with other bits of the universe outside the system. A closed system is one that is isolated from the rest of the universe – no matter or energy can flow in or out of the system. The universe itself could well be a closed system, though it is also possible that it is part of a larger multiverse and there could be energy flowing into or out of it.

Because of their origins, the laws of thermodynamics talk of things like temperature, heat and energy – which is fine, but is only half the picture, and not the bit where dice world comes into play. When these ideas were first put together, no one had a clear idea of what matter was made of. Atoms were nothing more than theoretical constructs. As we've seen, it wasn't until Einstein's work on Brownian motion in 1905 that atoms were taken at all seriously, and even for a number of years after that many scientists considered them useful concepts with no basis in reality.

Now, though, we know differently. All matter is made up of vast numbers of atoms. (In practice they will often be arranged in molecules, but we'll just talk about atoms for simplicity as exactly the same goes for either in terms of thermodynamics.) The concept of temperature established in the zeroth law is not some arbitrary property, but a measure of the kinetic energy of the atoms of a substance – in effect, a measure of how fast they are jiggling about. We can't realistically find out how much energy each individual atom has, but luckily this problem proved ideal for statistics. Rather than worry about each individual atom with its own energy, we think of the whole, varying from atom to atom according to a known distribution and so open to the siren song of statistics. Although the early practitioners weren't aware of it, thermodynamics is fundamentally a statistical concept. You can't get away from it.

Quantum mechanics tells us that atoms can only be in certain energy states, jumping between them in quantum leaps. When we look at a vast collection of atoms through the eyes of statistical thermodynamics, we see a distribution of different energy states. For a seriously cool object near to absolute zero there will be relatively few different states, with most near the 'ground state' – the lowest possible state. With hotter objects, the distribution will be more spread out across higher states as well as the ground state. This is not a normal distribution, but an exponential one. Statistically, temperature is a measure of the way those states are spread out.

The coldest possible temperature, absolute zero, represents a situation where all the atoms occupy the ground state. In any normal system, any particular atom (or molecule) may change state at any time, but the distribution corresponding to the temperature is the most likely one for the system as a whole to occupy.

When we look at the zeroth law from the point of view of a distribution of kinetic energies (and hence speeds) of atoms, it makes good sense. If two bodies are in contact and they both have the same distribution of energies, there is no reason why one should influence the other. Sometimes a high-speed atom in one body will bash into a lower-speed atom in the other body and a little energy will be transferred. But averaged out across the whole distribution, equivalent amounts will flow each way. It's an equilibrium.

It's worth also looking at the first law through a similar atomic lens. If we do work on a system by applying force to it, or we heat it, in both cases we are transferring energy, but in importantly different ways. To do work on atoms we effectively give them a shove in the same direction. This takes energy: to do work usually means moving something against a force that is trying to prevent the movement. If we heat up a collection of atoms we speed them up as they zoom along on their chaotically random collection of paths. Work is a collective shove in a single direction, heat is a set of shoves on each atom independently in the directions they happen to be travelling in.

The law of change

And so we come to the big one – the second law of thermodynamics. The one that tells us how the universe works, and how it's likely to end. This is because the second law tells how – and effectively why – things change. Everything from the evolution of the universe to the smallest aspect of a human life is essentially about how things change from moment to moment, and the second law is always there, silently steering the course.

The concept that tends to trip people up when they are trying to understand the second law is entropy, the property of a closed system that the law says will stay the same or rise. Entropy is often described as the amount of disorder in the system. The bigger the entropy, the more the disorder. Left to its own devices (which is another way of saying in a closed system) the entropy will only typically rise, or at best stay the same. If entropy was only that fuzzy concept of amount of disorder it would be very qualitative and not particularly useful in physics – but it can, as we will see, be quantified thanks to statistics.

To get a very simple picture of entropy, take a look at the letters on this page. As they stand they are in a low state of entropy – they are in a particular order that gives specific meaning to the page. If the letters weren't attached to the paper and you gave it a shake and jumbled them up, then the entropy would have increased as the letters went into a disordered state.

One way to compare the level of order or disorder is to compare the number of ways you could have the letters in the two states. There is only one combination of positions of those letters that would spell out the exact contents of this page. (In practice you could move quite a lot of the letters around and still transfer the same information, but that reflects how clever you are and your brain's ability to find patterns in what it is seeing. If an idiot computer was asked if it was the same information it would say 'No' if I just moved the full stop at the end of this sentence outside the bracket.) There are, however, many different ways to scramble up all the letters on this page – an enormous number. So a jumbled-up version of the letters has significantly higher entropy.

The 'closed system' bit of the second law is of intense importance here. The second law does not say that entropy will always increase in a system. Think about the Earth as a system. When you look at all the complex organisation not just of human technology but everything that goes into making living animals and plants, the Earth as a system clearly has much lower entropy than it had in the past when all the atoms and molecules were pretty randomly scattered about. Entropy on the Earth has decreased over time as more and more structures and patterns have been added.

I have seen books trying to counter the idea of evolution using this observation. They make use of this decrease in entropy as a sign that there has to be an active creator to 'break' the second law. But the second law only applies in a *closed* system. It's fine to consider the Earth a system, but it is anything but a closed system. Vast amounts of energy flow into it from the Sun all the time. Shove energy into a system and of course you can decrease entropy. Think of a fridge, for example. Remember the temperature-based version of the second law is that heat doesn't flow from a colder to a hotter body. Yet that's exactly what happens in a fridge. Heat flows from the inside of the fridge to the (hotter) outside. It can do that because, once more, the fridge is not a closed system. It's plugged into the mains. Energy flows into the fridge, and it takes that energy to be able to transfer heat the 'wrong' way.

Even the most obscure ways of producing order fit in with the second law of thermodynamics. Take, for example, self-patterning systems. A simple example of a self-patterning system would be a tray covered in a thick layer of wax. You put the tray at an angle and pour hot water on to the top edge. To begin with the hot water would run down the face of the wax in a highly disordered fashion. Soon, though, indentations would begin to melt into the wax. Once there is a set of pathways to run down, a fair amount of the hot water will follow the indentations, rather than skitter across the surface.

As more hot water runs down a particular route, the indentations will become bigger, carrying more of the water. Once the pattern is established, using it reinforces that pattern. The particular pattern you end up with has lower entropy than the ability to run all over the surface of the wax. You have, apparently without doing any work, reduced the entropy of the water/wax system. In reality, though, once more it took energy to make this happen. The water would be slowed down by contact with wax, using up some kinetic energy, and the water would cool, losing heat energy. As this and many other examples demonstrate, one way of looking at the second law of thermodynamics is TANSTAAFL. There ain't no such thing as a free lunch.

The particular example of running hot water down wax on a tray may seem a pretty obscure example. It's not something you do every day. (Well, it's not something *I* do every day.) However it is a very crude model of a natural self-patterning system that is very important to us all: a brain. As far as we can tell, the brain stores information in a kind of self-patterning system. The more particular links are used, the fatter they get, and the easier to use they become. This is good because it makes it easy to respond quickly to a familiar challenge – though it does mean that under significant stress we tend to lose creativity. And the brain, like the wax tray, also experiences TANSTAAFL. Using our brain takes a lot of power – around 20 per cent of the around 100 watts that the body uses at rest.

If we look at the version of the second law that says heat doesn't go from a colder to a hotter body, it seems common sense when we consider what is happening at the level of atoms or molecules. Heat, remember, produces the increase of the random kinetic energy of the atoms in an object jiggling about. If two bodies are in touch, so the jiggling atoms of one can bump into the jiggling atoms of the other, then the hotter body – the one where the atoms are moving faster – will pass on more of its jiggling to the atoms of the cooler body. There will also be a transfer of energy from cool to hot – because even the cool atoms are moving – but at a much lower rate, so the net effect is that heat flows from hotter to cooler.

Quantifying disorder

We've said that entropy is a measure of disorder, but also that as used in physics it is a quantitative measure. It's not just touchy-feely 'this seems more messy than that' – you can apply numbers to it. For this, too, we need to take an atomic or molecular view and have to introduce a quick equation, which is more painless than it looks. The equation is:

$$S = k \log W$$

S is the entropy (E was already used up for energy when this equation came along), k is a constant called Boltzmann's constant and W is the number of ways a system can be arranged to achieve the particular result (we'll worry about that 'log' in a minute). Think of the example of the letters on a page of this book. If you imagined there are a series of slots on the paper that you can put letters in (think of the old moveable type printing press), then it's easy to see that there is one way to arrange the letters to get a specific page, but by trying each letter in each slot you could (very slowly) work out W for randomly distributing all the letters and would get a much higher value.

Mostly entropy isn't about letters on a page but about stuff, and particularly the atoms or molecules that make that matter up. There again, in principle we can imagine different values for entropy for, say, a crystal where all the atoms have to slot into specific positions and a gas where they're bouncing all over the place. We couldn't do the sums exactly – and would have to resort to statistics to get anywhere – but it's entirely possible to see how entropy applies this way in theory.

A quick explanation of that 'log' bit. Fifty years ago I probably wouldn't need to do so, as most people would be familiar with logarithms from school, but as they have become less necessary for everyday calculations they have disappeared into obscurity as far as most of the world is concerned. A logarithm (abbreviated to 'log') started life as a way of making multiplication and division easier to do. It is based on an interesting observation about mathematical powers.

In science we often write a number like 10^6 – which is a compact way of saying 1 million: 1 with six zeros after it. It is 10 multiplied with itself six times, or $10 \times 10 \times 10 \times 10 \times 10 \times 10$. If I multiply a million by a thousand, I get a thousand million (a billion), which is 1 with 9 zeros after it. Now here's the interesting thing. If I multiply 10^6 (a million) by 10^3 (a thousand) I get 10^9 (a billion). Notice what has happened to the powers of the values I multiplied. They just added together. The logarithm of a number is just the power bit. So the logarithm of a million is 6. And to multiply two numbers I just add their logs. Adding is a lot easier than multiplying if you are dealing with more messy numbers – so logs are useful.

Logarithms crop up in the natural world too. When something like entropy varies with the log of a value, it means that a small change in the entropy can result in a big change in the value. When I just add 1 to the log of a million, taking it from 6 to 7, the value jumps from 1 million to 10 million. The bigger the value is to start with, the bigger the change an increment in the log will have.

There's one other distinction to be made. To keep my explanation of logs simple I used logs to the base 10. So if the logarithm is 1, the value is 10, a logarithm of 2 gives a value of 100 and so on. But I didn't have to. I could use base 3, for instance. So a logarithm of 1 would give a value of 3 and a log of 2 would give 9. A lot of logarithms in nature have a rather fiddly base called e, which is around 2.71828 – it's not nice and convenient, but like that other messy natural number pi, it's what we have to use to match the natural world. The 'log' in the entropy formula is such a 'natural log', based on e.

This formula $S = k \log W$ was first written down by the physicist Max Planck, but it very much owes its existence to the mathematician Ludwig Boltzmann after whom that constant is named and who produced most of the basic maths of statistical thermodynamics. In fact, the formula is carved into Boltzmann's tombstone. It's a fitting epitaph, as Boltzmann committed suicide suffering from depression. He may well have suffered from bipolar disorder, but if there was any reason for his depression it is often said that it was because of the lack of support for the concept of atoms, something that was fundamental to his statistical approach to thermodynamics.

Entropy plays dice

To the original creators of thermodynamics, with steam engines in mind, there was no doubt about the nature of the second law. It was a statement of fact: whenever there was a change in a closed system, entropy would rise. It was only if there was action from the outside that entropy could fall. However, in the dice world that is the true reality, certainty is removed.

All we can say in the real, statistical version of thermodynamics is that the vast majority of the time we expect that entropy will rise – but it doesn't have to be that way because randomness is at play. Think of a really simple system that is a closed rectangular box we can fill with a gas*. (It could be any gas – air, for example. But to keep it simple it might be easier to think of a single substance, like oxygen.) Our box is constructed so we can slide a divider down the middle to separate the box into two parts. Let's do that. We then fill the left-hand half with hot gas and the right-hand side with cooler gas. Then we take out the divider.

Over time the gas from the two halves will mix. Bearing in mind that temperature is just a measure of the average kinetic energy (and hence speed) of the gas molecules, the mixture will end up at an intermediate temperature between the two. Now what has happened here is that the entropy of the box as a system has increased. We had more order to begin with – hot on the left, cold on the right. Now the molecules are all over the place. We could work out the entropy value numerically with the Boltzmann equation, but it's obvious from the simple description that it's true.

---

* Pedants will point out this box is actually a rectangular parallelepiped, as a rectangle is two dimensional, but you know what I mean. A box the shape of a brick. With no dimples.

According to basic thermodynamics, that's it. End of story. Entropy will always increase. But because the reality is statistical, there is a small chance that spontaneously, if we wait long enough, the random motion of the molecules will result in the hotter molecules being on one side and the colder molecules being on the other side again. It's very, very unlikely. But it could happen with a predictable probability (this is classical randomness). For the moment this is just an interesting twist to the second law – but it will be significant when we start to think of one of the facts that emerges from the second law as applied to the universe as a whole. So significant, in fact, that it calls into question our understanding of time.

First, though, we need to see how a variant on that simple box of gas could baffle physicists for decades, thanks to the evil-sounding Maxwell's demon.

CHAPTER 14

Maxwell's demon

Before we meet the demon, I ought to take a moment to consider the dangers of simple models. This isn't an insult to those who work on catwalks, but a concern about the way physicists view the world. There's an old joke I've told in print before, but it bears repeating. A dietician, a geneticist and a physicist were arguing about how to produce the perfect racehorse. 'It's obvious,' said the dietician. 'Make sure you have a perfect diet for the animal and you will succeed.' The geneticist shook her head. 'It's all in the breeding,' she said. 'You've got to breed selectively for the right traits.' The physicist has been quiet so far. He slowly shakes his head. 'Look,' he says, 'let's imagine the horse is a sphere.'

That may not sound side-splitting but I can assure you it will reduce a room full of scientists to tears of hilarity. Physicists are infamous for working with very simple models of reality. We're back to the person searching for their keys and looking under a streetlamp, even though they don't think they lost them there. It's the only place they can see. Similarly, the real world is much too complex to produce detailed mathematics to describe it, so in physics we tend to work with simplified versions of reality. (Hands up who did physics problems at school saying '… ignoring friction'? In the real world you can hardly ever ignore friction.)

In some circumstances, such simplification to make a problem tractable is dangerous because you can draw conclusions about the model that don't have any significance for reality. Take a ridiculous example. I could build a robotic car that drives itself and the model I could use would be that the traffic on the road will be the same as it was yesterday. Exactly the same. I just record a journey, showing the car when to stop at red lights, when to avoid other cars and so on. And then I set it going in today's traffic. Carnage would ensue. But there are plenty of circumstances where we can use a simple model to understand something fundamental, as long as we are aware of the limitations of the model, and that's what we are going to do here.

Unmixing the mixture

So to the demon. We're going to stay with the box of gas introduced in the previous chapter, as there is more fun to be had with it*. A small extension to our simple model produces a paradox that had physicists baffled for years. We are going to start with a box that has had a good

---

* Until you discovered thermodynamics, you probably didn't realise that you could have fun with a box of gas. It used to be possible in the real world. When I was at school, a favourite demonstration of science teachers was to fill a metal container (a large coffee tin was the usual box used) with gas from the gas tap. The tin had holes punched in the top and bottom. The gas was lit at the top hole. As it burned, air flowed into the bottom of the tin until the mix of gas and air in the tin was just right, at which point it exploded, blowing the top off the tin. It's not generally done in schools any more because of the health and safety risk – so don't try it at home.

chance for the molecules to mix and balance out. There is a uniform mix of fast and slow molecules on both sides of the box.

Imagine we have a very small person, so small that he can see the individual molecules of gas passing back and forth in the box. And he has such a high metabolism that he can respond to the different molecules one by one as they pass by. He is in charge of a trap door in the divider that splits the box in half, a special trap door that it takes no energy to open and close. As a molecule approaches the (closed) door our little man takes a look at it. We'll call him a demon. If the molecule is moving left to right and it's fast, he opens the trap door and lets it through. If it's going slow he leaves the trap door shut. Similarly, if a molecule is moving right to left and it's slow, he opens the trap door. Fast molecules from right to left don't get through.

After a while, our demon will have many more fast molecules in the right-hand half of the box than in the left-hand half. He will have started with a mixed gas with a middling temperature throughout the box and will have ended up with cold gas in the left section and hot gas in the right. He has reduced the entropy, because the gas molecules are now more ordered than they were before, the reverse of the original action of opening the partition and letting them mix. But assuming he really can open and close the trap door with no energy used, then he really is a demon. He has defeated the second law of thermodynamics.

I've called the little creature a demon because that's what he's traditionally called – specifically Maxwell's demon*, named after the great Scottish physicist James Clerk Maxwell. Physics students rate Maxwell with the greats like Newton and Einstein, primarily because of his work on light, explaining how it was an interaction of magnetism and electricity – in fact it was Maxwell's work on this that inspired Einstein to come up with special relativity. But Maxwell was a versatile thinker who also took the first colour photograph and did important work on statistical thermodynamics. Maxwell didn't call the creature a demon – that was down to Lord Kelvin – but it was Maxwell who dreamed him up.

Maxwell's demon teeters on the edge of whether or not the model has started to become dangerous. We can't imagine a real creature capable of this act, or even a physical device (though they can be made if you provide some external energy). But the model is still useful in thinking about the nature of entropy and the second law. Although there is not total agreement on why Maxwell's demon doesn't break the second law, there are a number of arguments that make it unlikely he ever could. All of these rely on the fact that the demon has to be part of the system. If he's outside the system it's not a closed system, so the second law is irrelevant.

---

* There's no evidence that Maxwell's demon inspired Paul McCartney's 'Maxwell's Silver Hammer', but interestingly McCartney said the silver hammer was 'my analogy for when something goes wrong out of the blue, as it so often does …' – in effect the silver hammer in the Beatles song is chaotic randomness.

The simplest argument is that in the real world the demon can't have a trap door he can open and close without exerting energy. This is true, but the energy introduced into the system would not be sufficient to counter the decrease in entropy. What we then have to throw in is that the demon isn't allowed to use magic. He may be a demon, but he is working in the field of physics. Because of that, he can't just psychically 'know' how fast a molecule is travelling. He has to make a measurement, and that should involve exerting some energy – enough to make up for the decrease in entropy in the gas.

However, by the 1960s, it was realised that there are some means of measurement that can be undertaken without an increase in entropy in the demon to counter the decrease in the gas. The demon had a get-out clause. Or so it seemed. Because the demon's renewed ability to conquer the second law was dashed by a very strange effect. It couldn't get away with its trick because it would need to forget information, and to do that means exerting energy and an increase in entropy. What's more, this discovery depends on the relationship between entropy and information, a link that ties entropy into randomness and chance more firmly than ever.

Uniqueness is disorder

We've got the idea of entropy when applied to atoms or molecules as being the amount of disorder – but you can also think of entropy in terms of information. The more predictable the information, the lower the entropy. This will eventually lead us back to Maxwell's demon and his memory, but let's take a moment to explore further.

Think of a very basic form of information – a number with lots of decimal places. If you think of a number like ⅓, it's very predictable. As a decimal it is 0.3333 ... All I have to do to specify it is either say '0.3, then repeat the three' or 'divide 1 by 3'. Low entropy. But some numbers aren't so amenable.

Think of the square root of 2. It starts 1.41421356 ... and goes on without repeating itself. But it's fairly easy to calculate, and that calculation is, effectively, the shortest way to produce the number – it has a form of predictability. Longer than 0.3333 ..., but still predictable. Even worse is something like pi – 3.14159 ... Although visibly similar to the square root of two, the difference is that pi is transcendental. This means that while you can write down a formula to calculate it, you can't do it with a finite formula – it has to be in the form of an equation that goes on forever. For example $\pi/2 = \frac{2}{1} \times \frac{2}{3} \times \frac{4}{3} \times \frac{4}{5} \times \frac{6}{5}$ ... (see page 87) does calculate pi, but you will never complete the calculation. That might seem the extreme of entropy for a single number, but it's possible to go even further.

The ultimate for a high-entropy number, referred to as Omega, was dreamed up by American mathematician Greg Chaitin. Omega has no structure at all. You literally can't write down a way to calculate it. The only method to reproduce Omega is to copy it out digit by digit – it can't be compressed in any way. That makes it the definitive number as far as extreme entropy goes.

When we think of entropy in information terms, the early example of a page of this book makes sense too. There are various ways that the text in the book is predictable. There are the rules of English – so, for example, if you see a 'q' you know it will be immediately followed by a 'u'. There are lots of ways to compress the text in a piece of English, due to predictability. And if the page is to have its specific meaning, there is no real randomness to it. But the scrambled set of letters is a whole different ball game. Like Omega, there is no way to predict what will come next – the only way to reproduce the page is to copy it out letter by letter. It is truly random. So it has a lot higher entropy.

Back with Maxwell's demon, the demon has to store information to do the task that it has set itself. To decide whether a particular molecule is fast or slow needs an understanding of what the speed it measures means, and as it makes a measurement of each molecule it is collecting information. If the demon then erases the information, it's a feature of the entropy of information that performing that deletion takes energy – it increases the entropy of the system. Storing information doesn't need to take energy – but it has been proved that however much storage the demon has, it must eventually start to delete information to keep functioning and at that point entropy begins to go up.

Even now, the whole business of Maxwell's demon isn't entirely sorted out. There are some loopholes in the arguments at the detailed level. But it seems likely that were such a creature to exist, he could still only have a minor impact on the second law. Which is perhaps just as well, as this strange law of statistics and predictability also seems to be tightly linked with our concept of time.

The mystery of time

There is an awful lot we don't know about time. What time *is* provides an excellent example. You can read Stephen Hawking's *A Brief History of Time* from end to end and not get the slightest clue of what it is*. But one thing we can pragmatically agree on is that time has a clear direction, sometimes called the arrow of time. In space there is no special direction. The Earth might seem to give us one – towards the centre of the planet, which gives us the idea of up and down. But take away that local distortion and there is no special direction in any of the dimensions of space that stands out from any other.

Time is different, though. Whether or not it's possible to travel in time – and there is nothing in physics that makes this impossible – time has that inevitable pointer saying 'this way to the future.' As far as common sense is concerned, this is no surprise. That's how time is divided up – past, present and future. It's just how things are. Yet as far as the majority of physics is concerned, there is no good reason for this distinction between forwards and backwards in time.

Most physical activities are reversible. If I show you a movie of an object moving through space, there is no way of telling if I am running the video forwards or backwards. However, the second law of thermodynamics runs counter to this. The second law provides a distinct direction

---

* This is despite the introduction of Stephen Hawking's book claiming that recent breakthroughs in physics suggest answers to the question 'What is the nature of time?' The question is never addressed in the subsequent text.

in time. It tells us that entropy increases – but that is really shorthand for 'entropy increases with the passage of time'. In a sense it could be said that entropy sets up time's arrow, pointing the way to the future. Which way does time flow? What is the direction of the future? The direction in which entropy increases for a closed system.

This becomes clear when we return to the video camera. Take a video of a cup floating through space and it is impossible to tell how time is flowing. There is no change of entropy. Run the video backwards and it is not at all clear which version is running forwards in time. But have that cup smash into a handy asteroid and you will get a totally different picture. The cup shatters into many fragments as it collides. There's nothing strange about that. But run the movie backwards and all the bits of cup reassemble. That does look weird. Because entropy seems to be decreasing with time. The cup is going from disorder to order with nothing directing it. It's not natural.

I put my imaginary cup in space rather than just dropping it to break it, because dropping something on the Earth has another reason not to be reversible. Objects that have more mass than the air don't float upwards. If you watch a dropping cup carefully, it starts off slowly and accelerates under the pull of gravity, giving a clear indication of the direction of time. A falling cup isn't an effective testbed for the way entropy gives time a direction because the gravitational pull of the Earth makes things happen in a particular time sequence, which is a giveaway that is not present in open space.

The clockwork runs down

The other, rather dismal apparent implication of the second law of thermodynamics is that it is horribly true that 'This is the way the world ends / Not with a bang but a whimper.' The reality is on a much greater scale than T.S. Eliot's original intentions, which were probably to refer to Guy Fawkes and the plot to blow up the Houses of Parliament. If we look at the universe as a whole, we would expect its entropy to continue rising with time until it consists of total disorder.

The timescales for this to happen are immense – so it's not exactly something to lose any sleep over – but in case you find the idea of the universal onset of chaos distressing, there are a number of get-out clauses. Bearing in mind the statistical nature of the second law, it is possible that entropy will not triumph forever. And just as evolution depends on the long lifetime of the Earth to date, the universe could be expected to be around long enough for something *really* unlikely to happen.

When we think of very unlikely events, it's not uncommon to put their probability into context using the lifetime of the universe to date. So, for example, quantum theory tells us that it's entirely possible for every atom in a car to move five metres sideways, leaving the car standing outside the garage it was once parked in. But the chances of this happening are so low that it's traditional to say 'You would have to wait longer than the [current] lifetime of the universe for it to happen.'

We don't really know how the future of the universe will pan out. How can we? All we have to base our ideas on is the past – and it is entirely possible there will be events that can't be forecast from the past. We could well be like the Christmas turkey, basing it's future on classical randomness, where the reality is more a case of chaotic randomness.

However, in the scenario where the universe simply drones on forever getting more and more disordered until it is just a set of uninteresting random molecules, there is an interesting possibility to contemplate. Given that the second law is statistical, if we wait long enough, unlikely though it may seem, the hot and cold molecules in our box of gas will at some point spontaneously separate into the two sides of the box. With a vastly longer waiting time, in principle, similarly, all the molecules in the universe could revert to a more ordered state with planets and stars, rather than a disordered diffusion.

There is at least one problem with this picture. The nuclear fusion reaction that powers stars has a natural low point. In fusion, first hydrogen fuses to helium, then heavier elements are formed all the way up to iron. But there simply isn't enough energy to go beyond iron – it takes the explosive power of a supernova (and not every star will ever become a supernova) to produce the heavier elements. Although the process of producing heavier and heavier elements is partially reversible – so, for example, very heavy elements will decay radioactively, producing lighter elements – and there's plenty of hydrogen and helium in the universe that is not in stars at the moment, the second law doesn't give us a way to jump back to the early years of the universe, undoing all the nuclear fusion that has taken place. However, in principle, given the eternity of future of a gradually decaying universe, in principle enough order could spontaneously emerge and re-emerge to sustain many cycles of existence.

A perhaps more likely get-around of a universe that gradually becomes more and more disordered is to consider what has happened to the Earth since it formed. The Earth has become more and more ordered over time. This has been a side effect mostly of life on the planet, powered by the Sun. As we've seen, the key to defeating the inevitable onset of disorder is an external energy source. The Earth isn't a closed system. But can we assume that the universe is closed either? We don't know that there isn't an external source of energy that could re-impose order on the universe.

It is certainly true that some cosmologies allow for our universe, the universe we experience, not to be everything, but rather a small part of a larger whole. Whether you support the idea of a bubble multiverse, where our universe is just one of many, possibly infinitely many, inflating bubbles in a much bigger system, or the ekpyrotic model of the universe, where our universe is a three-dimensional 'brane' floating in a multidimensional space that occasionally collides with another brane, producing a new big bang and a fresh start, it is entirely possible that there could be an external source of energy that would enable the second law to be thwarted – at least in the locality of our universe.

When the Victorians came up with the laws of thermodynamics they had everyday, practical goals in mind. It was all a matter of getting a better understanding of steam engines. But with a statistical interpretation of the second law and the wider implications that our modern understanding of entropy bring, the actions of Maxwell's demon and his probabilistic world continue to impact us all.

They certainly give us room for thought when it comes to the age-old activity of predicting the future.

CHAPTER 15

Crystal balls and winning goats

The future may be an uncertain destination, but we have always had a burning desire to know what it is going to bring. In a sense all the statistical forecasting we do, even all of science that has a practical application, is about predicting the future. Science that looks to the past, that merely tabulates what has happened and what exists, is what physicist Ernest Rutherford was referring to when he commented archly that 'all science is either physics or stamp collecting'. For science to be more than a matter of cataloguing, it needs to make predictions.

People have been telling us about the future as far back as we have history, and no doubt seers and prophets did so well before there were any records. This wasn't just because it feels good to know what is coming, true though that may be. It was a reliable means of gaining power. If you can predict what is going to happen with any accuracy at all, you can claim to be able to influence the future – and that makes you powerful indeed, because no one is going to risk making you angry.

Some who claimed to see the future were indubitably making the whole thing up, using it as a mechanism to gain power – others no doubt genuinely believed they had the gift of prophecy, whether as a gift of a god or as a result of what they would have considered scientific means. By far the longest-surviving mechanism for predicting the future is astrology. Anyone with even a vague idea of science these days would not think of astrology as science – and astronomers get very huffy if you call what they do astrology – but it was indubitably an attempt to make the prediction of the future scientific. It doesn't class as scientific now because it isn't anywhere near our best guess – but in earlier times there was less data available to counter it.

It's in the stars

The broad concepts of astrology date back at least 3,500 years and have been found in most ancient cultures. It seems to have been a very reasonable thing to these early civilisations that the stars and planets could influence what happened on the Earth. There were two versions of astrology. One was divination – an attempt to predict the future. This was the predecessor of the modern horoscope that supposedly tells avid readers what will happen to them in times to come. Even in medieval times this type of astrology was frowned on as either religiously dubious or scientifically ridiculous.

The second version was more like an attempt to take a scientific view, to provide an explanation of natural physical events by observation. It happens to be wrong, just as the ancient Greek theory of the four elements was wrong, but that doesn't make it totally unscientific. By medieval times, this style of astrology was treated as one of the arms of astronomy. It made no attempt to predict the future, it merely suggested that the influence of the planets at the time of birth could have an effect on the child, just as we now might consider that some of the mother's actions while carrying the baby, like smoking, could have an effect on the child's development.

We now know just how little influence the planets have at the range they are and using the pathetically weak force of gravity*. The mass of the midwife and anyone else in the room would have more influence than the planets can. However, the divination version of astrology, the one we mostly see active today, does indeed claim to provide a kind of vision of the future for the individual. Even most astrologers would dismiss the inclusive bucket of a forecast given in a newspaper for all people with the same star sign as bunk, but they do believe that their detailed charts, taking into account just what was where in the heavens at the time of birth, are genuine predictors of the future. They do so in the face of weighty evidence that there is no basis for believing astrology to be true.

Leaving aside the inability to explain the mechanism – why the position of a planet when you are born should have an influence on your future – it has long been pointed out how many problems there are with the concept. The Roman senator Cicero who lived over 2,000 years ago commented, 'Did all the Romans who fell at Cannae[†] have the same horoscope? Yet all had one

---

* Gravity is 10^{39} times weaker than that other familiar force, electromagnetism. If you doubt that gravity is weak, think of the way a little magnetic toy sticks to your fridge. All that's holding it up is the electromagnetic force of that tiny magnet. Pulling it down is the gravitational force of the whole massive Earth. The magnet wins.

† Cannae was a battle in 216BC, when Hannibal's Carthaginians killed around 60,000 Roman soldiers.

and the same end.' In more recent days, a 1985 study published in *Nature* pitted 28 highly rated professional astrologers against statistics. They were asked to produce detailed natal charts for over 100 volunteers, and then to match these to personality profiles of the individuals. In each case the chart was compared with the correct individual's profile and two others. If the astrologers had guessed randomly, they might have been expected to get around ⅓ correct. The astrologers felt they should be able to match chart to profile in at least half of the cases. In practice the outcome was almost exactly ⅓.

If we discount astrology, probably the most famous attempt to predict the future since the Delphic oracle of ancient times was the work of Nostradamus. Writing in 16th-century Paris, Michel de Nostredame was an almost immediate hit with his collection of four-line segments of text, or quatrains, making enigmatic predictions of what was to come. In the years since, the *Prophecies of Nostradamus* has rarely been out of print and a whole industry has sprung up linking the European seer's predictions to everything from the rise of Hitler to the assassination of John F. Kennedy.

The problem with this kind of prediction – an issue that also comes up trying to match many a fortune teller's forecast to the real world – is that the prophecy is so vague that it would be strange indeed if it couldn't be linked after the event to *something*. If you have enough text, you can find a link to almost anything, provided you try hard enough. Others have analysed the Bible (and others *Winnie the Pooh*) and claim it predicts all kinds of events. It really isn't hard to see how Nostradamus's apparent overcoming of the dice world is nothing more than justification after the event. No one has ever used Nostradamus to successfully predict a major event before it has happened – only to match the prophecies to events with hindsight.

Future vision

For more than 100 years now, attempts have been made to make scientific studies of 'precognition' – the supposed ability to predict what is to come through pure mental activity, which is arguably one of the least easy to explain of the so-called psi abilities. Probably the best work on this has been done by psychologist Darryl Bem, who published a paper in 2011 claiming to have detected precognition in action. However this was a very different feat from a Nostradamus-style prophecy. Bem performed a number of tests where the participants had to anticipate what was going to happen and detected a small statistical variation in the results from the expected outcome.

So, for example, in one of the experiments, the students acting as guinea pigs sat in front of a computer screen and were shown two pairs of curtains on an otherwise blank background. Behind one curtain was going to be a blank wall, and behind the other would be a picture. The students had to anticipate what would happen in the future by guessing which curtain would have the picture behind it. To ensure this was precognition at work, the picture's position was not selected until after the choice had been made*.

---

* As with all such tests, it is not 100 per cent clear what it was that was being tested. If the results had been positive, the

If the students got significantly better than 50 per cent of their guesses right, they would seem to be predicting the future. Interestingly the results were pretty well 50:50 when random pictures were chosen, but when erotic images were used, the students did do better than chance levels. But only just. They got the answer right 53.1 per cent of the time – over the number of experiments performed, Bem estimates there to be a 1 in 100 chance of this happening without a cause. Not a truly significant figure, but interesting. Bem did perform other experiments, which collected together did have a genuinely significant outcome in the mathematical sense, though they still relied on such incremental differences from chance. It is hard not to think that these results were caused by statistical or operational errors rather than an ability to see into the future.

Precognition may not deliver a useful picture of the future, but of course weather forecasters do give us a very limited view into one aspect of the future. As we have seen, modern forecasts are impressively accurate over a couple of days, become so-so over five days and are frankly useless over about ten days. These forecasts are put together by running complex models of the weather systems, charting changes from hour to hour in a series of three-dimensional cells that cover the relevant area of the atmosphere. Complex though these models are,

---

experiment could just as easily have been testing for the students' ability to influence the outcome, to change where the image was going to appear on the screen, as it was to predict the future.

making use of the world's best supercomputers to crunch the numbers, they did a pretty poor job, thanks to the chaotic nature of weather systems, until the approach of taking an ensemble of different forecasts (see page 98) was brought in.

A simulated world

As we have seen throughout the book, other kinds of forecasting rely on having the right kind of randomness driving what is happening. It is all too easy for the actual systems we are attempting to predict to be far too complex in reality for us to be able to produce any kind of confident forecast. Sometimes we can use a basic, distribution-based approach to predict the future. That's what I'm doing if I say how likely it is, say, to toss five heads in a row ($\frac{1}{2} \times \frac{1}{2} \times \frac{1}{2} \times \frac{1}{2} \times \frac{1}{2}$ which is a 1 in 32 chance). However even if the aspect of reality we are trying to predict is based on such classical randomness, it can be difficult to make predictions as soon as the system gains any complexity. This is where simulation can prove the best tool.

Simulation involves building a model, often on a computer, of a simplified version of reality. At each point in the model where there is a random occurrence, we put a random number generator, then run the whole thing rather like playing a game of The Sims on a computer, but repeating the same scenario many times to get an average picture of how things will develop.

A simple kind of simulation that I used to build when I worked at an airline is a queuing simulator. What I would do on a computer is set up a series of virtual check-in desks. Each check-in agent would be programmed to take a random time to handle a customer, with that random value being picked from a distribution that matched typical check-in times in the real world. Then I would have another random number generator introducing passengers into the system. Again, those passengers would be picked randomly from a distribution that was similar to the typical arrival time of passengers – this could vary according to the time of day.

Where it got interesting was what happened between the passenger arriving and getting through the queue. You could allow each passenger to choose a particular desk to queue at, like queues that are usually found in supermarkets, or you could have a single queue and allocate the person at the front of the queue to the next available check-in, as often happens in banks. I also had other random number generators handling whether or not individuals would tend to swap queues if they saw a queue moving much faster than their own.

If I had attempted to work out average waiting times for passengers in a system like the check-in environment of an airport using a single probability based on a single distribution I wouldn't have got anywhere – but by running a simulation like this (and running it many times) I was able to get a much better picture of what was really likely to happen.

This approach dates back to just after the Second World War when scientists at the Los Alamos Laboratory in New Mexico were trying to work out how neutrons would pass through various materials and what was necessary to provide shielding from such nuclear radiation. They found straightforward statistics didn't get them anywhere and thought instead of running a repeated simulation and accumulating the results. As this was top-secret work it had to be given a code name, so it was called Monte Carlo after the casino. The term stuck, so this approach is often referred to as the Monte Carlo method or a Monte Carlo simulation.

In the early days, such simulations had to be carried out by hand, aided by mechanical calculators, so the process was very slow, but with the development of computers they have become powerful tools for analysing situations that are too complex to be dealt with by a single-probability distribution but are still driven by classical randomness. They are used in all branches of science and business, and even in maths where they can be used to solve some mathematical problems, though most mathematicians wince at their use because they don't like the idea of getting to information by repeated computer runs, preferring an abstract mathematical proof.

Cars and goats

Whatever tool we are using for forecasting, we need to be aware of some of the traps that our probability blindness sets for us. The best known of these traps is usually described in a form that is somewhat reminiscent of the curtains in Bem's precognition test. Although the Monty Hall problem has been widely described, it is worth revisiting because it demonstrates the immense gap between our natural response to probability and the forecasts that emerge from it.

The name of the problem dates back to a 1960s American TV game show hosted by the Canadian presenter Monty Hall. In the final segment of the show, a contestant was given the choice of opening one of three doors. Behind two of the doors were lesser prizes (often goats in variants of the game) and behind the third was a car. The contestant was asked to pick one of the doors. He or she would win whatever was behind the door. So far, so simple. With three doors and one winning choice (unless you like goats) there is a one in three chance of being a winner.

Now the mentally taxing bit. Once the contestant had chosen a door (without opening it), Monty Hall would open one of the other two doors and show the contestant a 'goat'. So that leaves two unopened doors, one of which will have a goat behind it, one a car. The contestant would then be asked to make a final choice between the two remaining unopened doors. The question is to predict what the best tactic should be. Would it be best if the contestant stayed with their initial choice? Would it be best to swap to the other unopened door? Or was it a 50:50 chance, so it didn't matter what the contestant did?

When writer Marilyn vos Savant presented this problem in her answers column in the magazine *Parade* in 1990, saying the best strategy was to switch to the other door, she was swamped with letters, some from maths professors, telling her that her answer was wrong. Almost everyone writing in was sure that the contestant had a 50 per cent chance of winning with either door. Two doors to choose between, one with a car, one with a goat. What else could it be but a 50:50 chance?

One correspondent wrote:

> I'll come straight to the point. In the ... question and
> answer, you blew it! ... Let me explain: if one door
> is shown to be a loser, that information changes the
> probability to ½. As a professional mathematician,
> I'm very concerned with the general public's lack
> of mathematical skills. Please help by confessing
> your error and, in the future, being more careful.

There was even one correspondent from the US Army
Research Institute who wrote, 'You're wrong, but look at
the positive side. If all those Ph.D.s [who wrote in to contradict you] were wrong, the country would be in very
serious trouble.'

Perhaps the country was. Because despite getting letters to the contrary from the deputy director of the Center
for Defense Information and from a research mathematical
statistician at the National Institutes of Health, vos Savant
was right. If the contestant sticks with the door he or she
chose, there is a 1 in 3 chance of winning. If a switch is
made to the other door, there is a 2 in 3 chance of winning.

It's important to understand that this really is correct.
It is easy enough to write a computer simulation of what
is happening, and many of us did when first faced with
the problem. It's harder to get your head around why the
chance isn't 50:50. I find it best to think of it like this.
Everyone can agree that to begin with, the contestant had
a 1 in 3 chance of being right and a 2 in 3 chance of being
wrong, because two doors have goats behind them and
one a car.

Let's say the doors are red, blue and green, and our contestant chooses the red door. Then we know there's a 2 in 3 chance the car is behind either the blue door or the green door. Monty Hall now opens one of those doors. He knows where the car is and would only ever open a door with a goat behind it. Let's say he opens the green door. So now we know that there's a 2 in 3 chance the car is behind the blue or green doors, *and* we know it's not behind the green door. So the contestant's best bet is to switch to the blue door. Two thirds of the time he or she will win the car. It will have been behind the red door just one third of the time.

Born on a Tuesday

Another prediction problem based on probability proved to be the second most controversial entry ever in vos Savant's column in *Parade*, and once again her readers were convinced that she had got things wrong. This one is so counter-intuitive that even those who find the Monty Hall problem quite straightforward struggle with it. The problem sounds trivial enough, and comes in the form of a statement for which we have to predict the probability. It reads 'I have two children. One is a boy born on a Tuesday. What is the probability that I have two boys?'

It sounds trivial. The Tuesday bit is just window dressing, so we are looking at 'I have two children, one a boy. What is the probability I have two boys?' So with one child a boy, surely there is 50 per cent chance that the other child is a boy and a 50 per cent chance it's a girl. Which makes the probability of having two boys 0.5, or 50 per cent. There's a one in two chance. But unfortunately that is not correct.

The reason we get confused is that when trying to imagine the situation we think of the 'first' child we come to being a boy, then look at the options for the second child being a boy. However the description of the situation would also work if the first child is a girl and the second child is a boy. The only way to be absolutely certain is to work through every possible combination. It's a trifle tedious, but it delivers the result:

	Child A	Child B
1.	Boy	Girl
2.	Boy	Boy
3.	Girl	Boy
4.	Girl	Girl

These are the four possible combinations, each equally possible. Of these, three are situations that match my initial statement 'I have two children, one is a boy'. In all but case 4, one of the children is a boy. But only one of those three combinations with a boy also makes the second child a boy. So the answer to 'I have two children, one a boy. What is the probability I have two boys?' is not 50 per cent, or one in two, it is one in three. This part of the problem is probably on a par with Monty Hall in the difficulty of getting your head around it. But there is a more fiendish part. We were wrong in discarding the Tuesday. Saying the boy was born on a Tuesday changes the probability.

To see this we need a much bigger table. It starts like this:

	Child A	Child B
1.	Boy (Mon)	Girl (Mon)
2.	Boy (Tue)	Girl (Mon)
3.	Boy (Wed)	Girl (Mon)
4.	Boy (Thu)	Girl (Mon)
...		
14.	Girl (Sun)	Girl (Mon)
15.	Boy (Mon)	Girl (Tue)
16.	Boy (Tue)	Girl (Tue)
...		

In total we have 196 entries in this table. We go through every single sex/day combination in the first column combined with a girl born on Monday (fourteen of them in all), then every single sex/day combination in the first column combined with a girl born on Tuesday (fourteen of these too) and so on until we have cycled through every option for the second child.

Now we need to know two things. How many of those pairs feature a boy born on a Tuesday (like item 2 above) and how many of *those* have a second boy? We are going to have one combination of child A as a boy born on Tuesday with every possible child B – fourteen of those, plus thirteen other combinations where child B was a boy born on Tuesday, but child A wasn't (we have already counted the instance were both child A and child B are a boy born on Tuesday). So there are 27 rows that match our circumstance of having a boy born on Tuesday.

We now need to pin down how many of those rows had two boys. The first set of fourteen all had a boy as child 1, and half of those – seven also had a boy as child 2. Of the thirteen additional rows where child B was a boy born on a Tuesday, six would have child A also a boy. So of the 27 rows with a boy born on a Tuesday, thirteen of them have a second boy. The answer to 'I have two children. One is a boy born on a Tuesday. What is the probability I have two boys?' is 13 in 27 – almost, but not quite, one in two.

This really upsets common sense, if you are still daring to make use of such a fragile mechanism after all the random world has thrown at us. Just by specifying the day on which one of the children was born we change the probability of both children being boys from one in three to 13 in 27. Yet our minds rebel at this. Surely we could have chosen any day? The only way I can see to make some sense of this is to point out that in any particular real circumstance, you can't choose that day at random; it is extra information that depends on the circumstance. The boy will have been born on a particular day and the result of that is that it cuts down the options, just like Monty Hall did when he opened a door and showed a goat. It's just that the reality is harder to accept in this example.

Confusing though the Monty Hall problem and the children problems are, they follow nice, easy statistics – on average, for instance, two times out of three you will get the car if you changed door in the Monty Hall problem. But most of the things we experience in life don't have such clearly predictable probabilities. The real world is usually much more complex and less susceptible to analysis by conventional statistics. Luckily, there is another way of looking at statistics that does enable numbers to be used to make predictions where what is known is much more fuzzy. It's called Bayesian statistics, and for most of its life it has been severely frowned upon by the great and good of the statistical world, but it is an approach that is now much more commonly used and more widely accepted.

CHAPTER 16

The Reverend Bayes and the golden retriever

The statistical old fogies who have a problem with the Bayesian approach complain that it is too subjective – but as we will see, the whole point of the approach is to get a handle on problems where that level of subjectivity is necessary. The power of going Bayesian is that it will work whether we are dealing with classical randomness, chaotic randomness – or no randomness at all. The downside is that in some circumstances it won't be as accurate as the traditional approach to probability (sometimes given the label 'frequentist', because this method depends on making predictions about events that happen with a known frequency).

The rather odd, if mellifluous, name 'Bayesian' comes from the man who started it all, the mathematician and church minister Thomas Bayes, who lived in the first half of the 18th century. Bayes never actually published a practical version of his method, but he did describe the concept with a very specific case of it being used, while the final, more practical version was derived from his notes after his death.

In effect, to use Bayesian probability, you first take a best guess at the chances of something occurring and then improve on that guess given any subsequent information. As much as possible, the inputs should be facts, but the process works with the best data you have available

– a much more likely situation in the real world than having the perfect accuracy assumed by conventional statistics. The distinction from traditional probability is that instead of looking at how often things should crop up as predicted by some distribution, we are looking for our best knowledge on what is happening. The mechanism of Bayesian probability is obfuscated by unnecessarily complex (and rather quaint) terminology, referring to 'priors' and 'posteriors', but what is going on is often actually quite simple.

The rest of this chapter is going to take the form of an experiment. We are going to make a prediction using Bayesian statistics. As it happens there is no secret to the answer to the particular question I am going to use – I know what the answer is in advance. But I didn't *need* to know it. I will use some equations part way through this process, but there will be a chance to skip over them. However I would encourage you to check them out. The maths is really very trivial and this is a remarkably powerful technique that you may well find useful in the future.

What we are about to do is to make prediction possible given limited information. We will start with the basic information that we *do* have and try to improve the outcome until we have the best guess we can make given the evidence available. This is about as good as prediction can ever be if we don't have enough knowledge to apply traditional statistics usefully.

The case of the informative mug

So let's take an actual example of Bayesian probability at work. I have a dog. (This is true.) What is the probability that my dog is a golden retriever? Unless you know me, your immediate reaction might be 'I haven't got a clue' but we can do a bit of digging and start to get to a sensible guess. From the Kennel Club registration statistics, around 6 per cent of dogs registered in the UK in 2006 were golden retrievers. This statistic is pretty vague, as it is only for registered (i.e. pedigree) dogs, so the actual percentage is likely to be lower, but golden retrievers have also increased in popularity over time, so that might balance out. Anyway, it's the best we can do.

So our starting point, with no other knowledge, is to assume that there is a six in 100 or 6 per cent chance that I have a golden retriever. It's not an accurate value, but it is the best we can do with no further information. It's better than nothing. If we now get hold of some extra information, Bayes allows us to add in that extra information and see how it changes our prediction. In this particular case (this also is true) it happens that as I am typing I have a mug on my desk with a picture of a golden retriever on it. Could that help us get our probability estimate more accurate? Let's see how that information changes things.

We have to be a bit creative at this point. I would guess that the chances of owning such a mug if you have a golden retriever are 50 per cent, but the chances of owning such a mug if you don't have a golden retriever are just 1 per cent. That's the subjective bit. I don't know those numbers – though I could research them and get better values. If I have that information, Bayes' theorem gives me a method of improving on my original 6 per cent estimate. If you get a pain between the eyes when faced with an equation or two you can skip the next page or so – but I recommend carrying on and taking it slowly. This genuinely is a lot simpler than it looks at first sight.

Bayes' theorem says:

$$P(A \mid B) = \frac{P(B \mid A) \times P(A)}{P(B)}$$

In English, that says that the probability of A, given B, is equal to the probability of B given A times the probability of A, divided by the probability of B. With these kind of equations P (something) means the probability of something, and P (something | something else) means the probability of something given something else. It's no more complicated than that.

Let's use my example to make things clearer. We'll turn A into G for golden retriever, and B into M for mug. So P (G), for example, means 'probability of a golden retriever' and P (G | M) means the probability of a golden retriever given a mug with a golden retriever on it.

So the equation becomes

$$P(G \mid M) = \frac{P(M \mid G) \times P(G)}{P(M)}$$

It says the probability of having a golden retriever, given that I have the mug, is the probability of having the mug if I have a golden retriever times the probability of having a golden retriever and divided by the probability of having the mug.

Now we need to fill in some values. First the top bit of the equation. We said the probability of having the mug if I have a golden retriever is 50 per cent – 0.5. And the probability of having a golden retriever is 0.06. So the top bit – P (M | G) × P (G) – is 0.5 × 0.06 = 0.03.

We also need the probability of having the mug or P (M) for the bottom of the equation. This has two components. The chance I have the mug and do have a golden retriever, and the chance I have the mug and don't have a golden retriever. We can write that out as:

$$P(M) = P(M \mid G) \times P(G) + P(M \mid g) \times P(g)$$

where 'g' means 'not having a golden retriever'. It's the probability of having a mug given a golden retriever, times the probability of having a golden retriever, *plus* the probability of having a mug given no golden retriever, times the probability of no golden retriever.

So P(M) = 0.5 × 0.06 + 0.01 × 0.94 (because given that the chance I have a golden retriever is 0.06, the chance I *don't* have a golden retriever is 0.94).

A quick rattle on the calculator tells me P(M) is 0.0394

** REJOIN US HERE ** if you don't like equations.

So going on my guess about the significance of the mug, the chance of my having a golden retriever given that I own the mug is 0.03/0.0394 – which is around 0.76 or 76 per cent.

Something remarkable has happened. Remember, my original prediction for the chances of my having a golden retriever, only knowing that I have a dog, was a 6 per cent chance. Now, given that I have that mug on my desk, the chance has risen to 76 per cent. It is beginning to look significantly more likely. And I can reveal that Bayes was right – I do have a golden retriever.

Good guesses are better than nothing

Those who don't like Bayesian statistics will start mumbling and muttering at this point. Yes, they will say, but the information you used to modify the original value was guesswork. We don't know how many people who have golden retrievers have mugs like that, nor for that matter how many people who don't have golden retrievers have mugs like that. It was those figures of 50 per cent and 1 per cent respectively that made the whole thing work. But these seemed *reasonable* figures to me. And they may be the best I can do.

Ideally I would improve on those values. I just came up with those numbers off the top of my head. A quick and dirty way to enhance them a little would be to ask a few other people what their best guesses of those two figures are. This is the so-called 'wisdom of crowds'. It is much less mysterious and mystical than it sounds. Almost always, if you have a single guess for a value (or a single poll if you take an opinion poll), then combine that with a range of others, you will improve the quality of the guess. This is because the chances are that the first person you ask (or first poll you take) won't be the closest to the actual value, but averaging over a range will include several who are closer, and will smooth out any extreme ideas, as one individual might not be representative.

Of course, there is always the danger that bringing in others will make things worse. Say, for instance, I was asking people's opinion of the outcome of the Monty Hall problem (see page 231) and the first person I asked was an expert in probability, who would get the answer right, while most of the other people I asked were unfamiliar with the problem and would get the answer wrong. Then the crowd would make things worse. But in most cases, if we ask people to take a guess at information, the answer is not so clear-cut. Certainly in the golden retriever problem I don't think anyone would have special expertise and so it is a valid assumption that asking a group would improve my estimate.

We saw earlier (page 107) that there is always a danger when asking a group of people for their opinion that the way you select that group may have an influence on the outcome. If, for example, you ask people at a political party conference their opinion of the leader of an opposing party, the chances are that the responses you get will not be typical of the population at large. I decided to take a straw poll on Facebook, which was certainly not representative of the whole population. My sample was only from people who are friends with me on Facebook. These people have to have a certain level of technology just to be on Facebook, and among my friends there will be more people living in the area where I live, and more people with a scientific background than there are in the population at large. But I can make a judgement that this won't produce a significant bias in this instance.

We will come on to what I found out from my straw poll in a moment, but let's just see how we could have improved things even more. There is quite good evidence that if you ask a group of people for a guess at a piece of information where they have no expertise, then show the group the results and ask them to modify their first guess, they will produce a better outcome than if you just went with their first guess. This so called 'Delphi' process seems to work a fair percentage of the time to slightly improve values. I didn't want to hassle my information providers by asking them twice, so I don't have such a 'Delphi' group, but there is a way I can approximate to the Delphi effect.

The tendency in the Delphi process is to eliminate extreme outliers – the highest and lowest values. If people originally guessed extreme values, they may well look at the crowd and think 'if everyone else is around there, why am I so far out?' So I can approximate to a Delphi outcome by eliminating the top and bottom 10 per cent – where I'm likely to find the extreme values that will distort the average, like Bill Gates in a room – and *then* take an average. I will be removing any freak values and homing in on the general feeling of the group. This is only a test to see how much the result will change. I needn't take the new value – I can stick with my original one – but by doing this I get a feel for how much the result is being influenced by extreme guessers who could be in a world of their own.

When I take a wider set of opinions by doing a poll of some kind I am obviously still only combining guesses. I could get closer still to the truth by doing a survey of mug owners, not asking for opinions but asking actual people for facts on how their ownership of golden retriever mugs corresponded with dog ownership. Depending on my sample size, I could come close to obtaining real values for those two percentages. But part of the point of this exercise is to demonstrate how powerful Bayesian statistics are for making a prediction given limited information. Bayes won't necessarily produce the correct probabilities, but it will give you a best guess, given the information you have.

So what was the outcome? In response to asking for help on Facebook I got 31 replies to add to my own. Averaged across those replies, the probability of my having a golden retriever given that I have the mug is 56 per cent. When I knocked off the bottom and top 10 per cent of values this only shifted to 57 per cent. This was despite having a couple of responses that I found quite surprising. One individual (who, as it happens, I know owns a golden retriever) thought it was equally likely someone would have the mug whether or not they owned that breed of dog.

Even more extreme were two individuals who thought that someone would be more likely to have the mug if they *didn't* own a golden retriever. Their argument was that it was an aspirational dog, so quite a lot of people would have the mugs because they wished that they could own the dog. These two were, in some ways, forming a population on their own, as they both work in the same office and didn't see what everyone else was saying.

Mr Bayes's oracle

So how well do we do at making a prediction thanks to Bayes? What can we learn from the exercise? A first point is that there seems little doubt that having that mug significantly increases the chances that I do own a golden retriever. In all but the three responses described above, the chances shot up by at least a factor of three, and in many cases a factor of ten or more. The outcome is not definitive. Getting 56 or 57 per cent likelihood is not a convincing enough percentage to be sure about what the truth is. But what I can say is that this is our best guess, given the information we have.

If I were forced to bet on the outcome, for example, the 56 per cent figure gives me guidance that I ought to put my money on having a golden retriever. I might not win in over 40 per cent of cases, but it is still the best guess I have, given the information presented. And why would I go for anything other than the best guess?

Remember, this is what science is doing all the time. When we say that, for instance, the universe began 13.7 billion years ago in the big bang, we are not stating a fact. Science is not about absolute facts. I can never prove that the universe began this way. Evidence could come along that disproves the theory. It only takes one certain contradictory fact to kill a theory. But all we can say is that this is our best theory given the data that we have at the moment. Sometimes that might be a 99 per cent certain best guess. At other times, just like the golden retriever mug, it might be just a 57 per cent certain best guess. But it's still the best we have right now.

Dark matter would be a good example of this. One of the mysteries of cosmology is where all the mass in the universe comes from. When something like a galaxy is spinning round, we can judge from the way it sticks together (or flies apart) just how much mass there is in it. If we add up all the visible stuff in a galaxy, there is far too little to keep it all together, so cosmologists speculate that there must be 'dark matter' – extra dollops of massive stuff we can't detect.

At the time of writing, an alternative theory to dark matter, called MOND (a contraction of MOdified Newtonian Dynamics), is gaining support. This theory has been around a while, but has been reinforced by recent data. MOND says that there is no dark matter, but rather that Newton's formula for the gravitational force has to be modified slightly when dealing with something as big as a galaxy – a variation that is entirely possible. Dark matter is still our best-supported theory at the moment. But right now, like the golden retriever, it is probably only around a 57 per cent best guess.

Bayesian statistics has proved to be our best tool for predicting the future in a real, messy world that isn't driven by neat classical randomness that handily fits a distribution. You don't need Bayesian statistics if you are trying to predict the outcome of a toss of a coin or the spin of a roulette wheel, but the vast majority of real-world situations are not like that. When in chapter 9 we looked at the probability of a medical test having a false result, we were using Bayesian statistics. The method works by recognising that the information we have is limited. Rather than simply trusting that information, it factors in the chance of getting things wrong. In the case of the medical test we knew the error rates – often we have to make an estimate. This is the closest we get to a functioning crystal ball. But it certainly isn't a way of predicting what human beings will actually do in any particular circumstance.

Is there a way to do this? Does a deterministic view of the universe – the mechanical universe of Newton and Laplace – mean that all our actions are pre-ordained? Or does the randomness at the heart of the universe mean that nothing can ever be predicted? These are essential considerations if we are to take on one of the fundamental questions about the nature of humanity. Do we have free will?

CHAPTER 17

Free will?

For centuries, human beings have struggled with the concept of free will, and never more so than once it was possible to imagine that we lived in a deterministic, clockwork universe. How could anyone ever make a real choice if everything was determined in advance because each action at the atomic level led to the next? In the inexorable chain of events from beginning to end of the universe, Newton and Laplace left no place for any true decision making. Perhaps the ultimate impact of randomness on our lives is to leave some wiggle room for free will, if we feel that it is necessary.

When statistical predictions first became commonplace there was a concern that the very existence of such forecasts suggested an absence of free will. Let's say that statistics showed that one in 100,000 Londoners was a murderer. Or that one in 1,000 would suffer from a particular disease. It seemed almost as if these numbers were taking away individual freedom by looming over us like the hand of fate. How could people truly have free will if a certain proportion of them were doomed to commit a crime or to suffer from an illness?

This concern really doesn't make a lot of sense. To think that the numbers force us to behave in a particular way is similar to the gambler's fallacy that says that, for instance, if a roulette wheel has just come up with a red four times in a row it is more likely to come up with a black next time. In reality, the wheel has no memory. The next spin will have an equal chance of coming up red or black (if it's a fair wheel). There is no magic force impelling the next turn of the wheel towards a particular colour to pull the total towards the mean. Similarly the fact that other people aren't murderers doesn't make it more likely that you are a murderer.

The inverse of the gambler's fallacy is also seen in sport. It comes across in the idea that a sportsperson or team is 'coming into form' or 'building up momentum' with a run of wins, in basketball given the dramatic description that someone has a 'hot hand'. This last example was actually studied in 1985 by psychologist Thomas Gilovich. He found that 91 per cent of basketball fans believed that someone has a better chance of making a shot if he or she has just succeeded two or three times than if the shot was preceded by a couple of misses.

Many sports fans (and many sports players) think that a run of a certain outcome makes it more likely that this outcome will be repeated. It sort of feels right because you think that the player is on form if they've done well – or perhaps that they will be put under a lot of psychological pressure if they have failed the last few times. However, when Gilovich studied actual shots in one team for a season he found that players who had just succeeded with between one and three shots were no more likely to succeed than players who had just failed for one, two or three shots.

My lucky numbers

Another example of the way that a misunderstanding of statistics comes into gambling is the obsession with statistics in a lottery in terms of which numbers have come up most often – which could either be used to suggest that 1) if particular numbers haven't come up as often as the rest, they are 'due', or 2) that if certain numbers come up more often than most, they are 'lucky'. There is even one website for the Irish Lottery that comments, 'To improve your chances of winning the Irish Lottery we have compiled a stats section based on previous draws.' And that would help how exactly?

The UK National Lottery kindly provides a frequency table for the main Lotto game. From this, we can see that at the time of writing, 38 and 44 were the most frequently drawn numbers, coming up 241 times, where 20 seemed to be a particularly unlucky number, only having made it out 171 times. This might seem too wide a variance for the balls to be truly random, and if you look at how often the different frequencies came up it can look a little random.

But this is misleading, because there are too few occurrences of any particular frequency to expect a nice, neat distribution. If we divide the frequencies into eight different ranges, allocating the balls to one of eight buckets, we start to get something that is considerably closer to a normal distribution, allowing for the fact we are still dealing with small numbers. When the various values are plotted, we do find that most balls fall around the average number of draws and a few are off in the tails. What has happened in the past is no guide to the future.

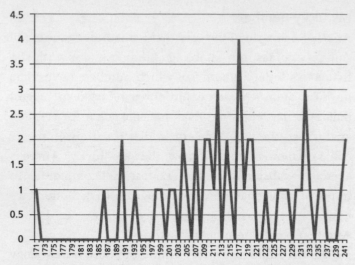

Occurrences of frequencies of ball selection in Lotto draws

All we know is that we will expect the overall distribution to stay roughly normal – but 38 and 44 are not 'lucky' and neither is 20 'unlucky' nor more likely to be drawn in the future than any other ball because somehow it is due a turn. Even the Lotto people have a rather bizarre view of this, as they specify the most 'overdue' numbers (at the time of writing, 47 hasn't been drawn for 188 days) – somehow suggesting that this makes them more likely to come up next time.

This doesn't mean that statistics can't be helpful in improving your chances in the lottery, just not in the way you might imagine. Statistics can't tell us anything about which balls are going to be drawn on any particular occasion. But they can help maximise your winnings if you are lucky enough to get the right numbers. Most lotteries have a prize pot that is divided among the people who

Frequencies of ball selection in Lotto draw in buckets

get all the numbers right. The more people who win on the same draw, the less cash you get.

To maximise your winnings, you need to make sure that you have a selection of numbers that is less likely to be chosen by other players. One thing to avoid is having all the numbers in your selection under 32. This is because many people use birthdays and anniversaries to provide their selection, and assuming they aren't also including years (unlikely in the range of numbers available in most lotteries), they will only choose numbers between 1 and 31. It might also be worth avoiding 38 and 44 in case people choose these thinking they are 'lucky'. Surprisingly, given how unlikely it feels, the sequence 1 2 3 4 5 6 is also quite popular. By having a near-random set of numbers that include one or more bigger than 31 you can somewhat improve your chances of winning a greater share of the pot.

Let's go back to my statistic about London murderers and imagine it was true (it's not, I made it up). One in

100,000 Londoners is a murderer. We go out and select 100,000 people at random, including you – if you don't live in London, we'll transport you there first. We go through the room one by one and by some magic means establish if each person is a murderer. We establish that the other 99,999 aren't. Does this change you in any way? If there is one murderer in every 100,000 people and the other 99,999 are innocent, does this make you guilty? Of course not.

There is no influence on an individual from the statistics. It's a one-way street. The people produce the statistics, but the statistics don't change the people. (At least not in the sense we are thinking here. Reading statistics can, of course, influence how people behave.) It's not that the statistics drain away free will, but rather that the collective actions of free will produce the statistics.

For that matter, we need to bear in mind the direction of time. Accurate statistics are always backward looking. They tell us what has happened. As soon as we use statistics to predict the future we have to assume that the appropriate distributions apply and that events are not connected to each other. This is certainly a reasonable assumption in many circumstances if, say, we are using statistics to predict the actions of many billions of molecules in a gas. But when we deal with a group of people it is all too easy for chaos to enter the mix.

Interestingly, and perhaps worryingly, we often see a sort of statistical fatalism coming into the way that courts take into account extenuating circumstances when someone has been convicted of a crime. The argument used may well be that '40 per cent of individuals with the accused person's background commit this kind of crime, a far higher percentage than for those who don't have this background.' We are asked to make the leap from the statistic to concluding that the background is a contributory factor to this individual committing the crime. And yet this is like assuming that a black number will follow a string of reds in roulette. Without specific evidence for a causal link for this particular individual, we are giving statistics more power than they deserve.

There is always an escape route from the apparent iron fist of statistics. In 1994, for the first time since records began, no eight-year-old girls died in Sweden. I am sure there are other occurrences of this kind, this just happens to be the one I've come across. Based on statistical evidence there should have been fatalities of eight-year-old girls that year. But the fact is none died. It was probably a result of both the steady increase in quality of healthcare and relatively small numbers, meaning that a year really isn't long enough a period to get a good prediction. But there is always a way to escape the idea that statistical predictions in some way limit what will happen. They don't, full stop.

More dangerous to the concept of free will is the idea of the brain as a mechanical device, a meat machine. We know that there are various factors that will influence our decision-making over which we have no control. There are genetic factors and medical ones. Brain tumours or brain damage can result in drastic changes in behaviour, sometimes leading to violence and other criminal acts. As people get older, diseases like Alzheimer's can effectively strip away the person we know and with it a lot of their free will.

I had no choice

It might seem that worrying about free will is just an exercise for philosophers who have far too much time on their hands. However, this is far more than an opportunity to indulge in abstract thinking. The existence of free will is a central tenet of the way our judicial system currently works. We assume that in almost all cases individuals do what they do out of choice that is enabled by free will. Occasionally we will reduce their liability when they are considered to be acting 'when the balance of mind is disturbed', but usually we expect people to take responsibility for their actions.

Take away free will entirely and there will inevitably be a defence that you shouldn't punish someone for anything that they do, as they had no choice. You may still incarcerate them to reduce the chance of them re-offending, but punishment and deterrence become meaningless concepts. The concepts underlying the legal system assume that alleged criminals are rational people who make all their decisions consciously, and so should always be culpable if they do something wrong.

The extreme downside of this comes through in the modern obsession with blame. In medieval times, unless magic was considered to play a part, when something went wrong it was assumed to be an act of God and people got on with their lives. Now, when anything significant goes wrong we assume that there is fault, usually both in individuals and systems. We want someone to take the blame. We can't accept that accidents will happen. Of course, some accidents are preventable, and sometimes the reason they aren't prevented is because of negligence. But the majority of accidents do not fall into this category. We shouldn't always be looking for a conscious actor to take the rap.

We only have to look at the realities of life faced by some individuals to see that the assumption of free choice of action is limited. Everyone occasionally makes an involuntary motion, one that has no conscious exertion to make it happen, but for some the experience is common. If you suffer from Tourette's syndrome, for instance, you will experience movements and even say things without any conscious effort. For the rest of us it can be difficult to understand that this is the case. But not only is it true, all of us have an even more dramatic break between our conscious will and our body's actions.

Back in the 1960s, American scientist Benjamin Libet discovered something remarkable about the way we make a conscious decision to do something basic like move a finger. Test subjects reported being aware of the intention to move about a quarter of a second before the action took place. But by monitoring brain activity using an EEG machine, Libet found that the brain kicked into action a good second before the individuals were aware of the intention to move. The unconscious brain starts the activity, *then* we become aware of the intention.

Some have argued that there are problems with Libet's approach as it was dependent on experimental subjects reporting the point in time when they felt they made the conscious decision to act. It is hard to see how this could be avoided, but it is a potential weakness in the experiment. Others have suggested that what is being measured looks like the precursor to a decision but is actually just spontaneous brain activity. Yet if Libet's theory is correct, the implication is quite simple. Our conscious decision process is more a rationalisation after the fact than a true act of free will.

It is possible to argue (as Libet did) that we still have time to veto the action in the quarter of a second of awareness we have before the physical movement occurs. More recent research has shown this to be true. Test subjects' brains showed signs of readiness to act whether or not they then actually went ahead with the action. But this is still an unsettling concept.

It would be interesting to really test any philosopher who denies the existence of free will as to what they really believe about themselves in their spare time. My suspicion is that after a hard day's pontificating they go home and know perfectly well that when they are deciding to drink a chardonnay rather than a claret, or choosing to listen to a particular piece of music, they believe that they are exerting a form of free will. But I'm not sure they would admit it.

A spanner in the clockwork

The random nature of the universe doesn't magically restore free will, but it does perhaps modify our view of a deterministic life in which everything is pre-written and we just go through the motions. Chaotic randomness is still deterministic, but makes it impossible to predict outcomes, cutting the ground away from under the tempting ideas behind Isaac Asimov's classic *Foundation* series of books.

In these stories, originally written in the 1950s, Asimov imagined a future branch of mathematics that was so sophisticated that it could be used to predict the future. Called psychohistory, it featured an intensely complex mathematical model of the galactic empire in which the stories are set. By careful manipulation of this mathematics, the Foundation could prepare for the inevitable downfall of the empire and plan for its rise.

In the books, the Foundation's mastermind, Hari Seldon, appears from time to time as a projection, giving uncanny predictions of what is happening at the moment. Eventually, though, things go wrong. The prediction begins to stray further and further from reality. The reason given is the rise of a man known as the Mule, who is a mutant and as such is somehow outside the predictive capability of the mathematics.

Although *Foundation* and its sequels first appeared before chaos theory was developed, it perhaps should have been obvious that the behaviour of human beings, let alone a whole intergalactic empire, was beyond prediction. The systems involved are just too complex and time and again we get the situation where small changes in initial conditions make huge differences down the line. Asimov assumed that just as statistical mechanics enables us to ignore the individual behaviour of atoms to get an overall picture of how a gas behaves, so statistical psychology – psychohistory – would make it possible to predict the behaviour of humanity as a whole. But the whole foundation of *Foundation* was wrong.

This chaotic randomness of human existence may not restore free will, but it means that we can't say with any certainty what the outcome of many of our acts will be, so we don't feel that our existence is permanently working along predictable tram tracks that guide us from beginning to end of our lives.

Then there is true randomness at the level of quantum particles. Everything we know is built on quantum particles. All matter, all forces. Everything. And every one of those quantum particles is just as weird and random in its actions as the decaying atom deciding the fate of Schrödinger's cat. So at a fundamental level there is no determinism. This doesn't give us free will per se. We have no control over the way those random behaviours play out. But at least it does mean that, even at the absolute level, the tram tracks don't exist. Stuff happens, and it's not pre-ordained. It might be a prediction for anarchy and pandemonium rather than rational, thinking action – but we aren't following a script. We may be puppets, but no one is pulling the strings.

What does this say for our legal system? If you accept that free will is a dubious concept, it needs a radical overhaul. What it doesn't mean is that there is a get-out clause for everyone who commits a crime, because they had no choice about it. We still need to prevent crime from being committed, but the focus has to move away from obsession with blame, which assumes free will on the part of an individual in almost all cases. Look at the case of Anders Behring Breivik who committed mass murder, killing 69 people in Norway in 2011. There was much debate over whether anyone could do this and truly be sane, as Breivik was declared to be.

It is hard to argue that this was a rational decision on Breivik's part. An act of reasoned free will by a sane person. And yet it cannot be simply ignored and treated as the symptom of an illness. There really is no way that this can be seen as anything other than the result of the nature of Breivik's brain, something he had no choice over. But at the same time, society needs to act to protect itself. Arguably, the only real meaningful prison sentence would be to lock someone away for the minimum time that would keep the risk of reoffending to a predetermined level. It's not an emotion-led approach – and not at all easy to calculate those risks – but it makes more sense than trying to decide if an individual was exercising free will or not when arguably we are all at the mercy of the chemical and physical composition of our brains.

The random deity

As well as considering the impact of randomness on our own free will, philosophers have also regularly explored the implications for the existence of a god. As we have seen, Laplace thought that his deterministic universe did not require a god. It certainly didn't need one to keep things going once they were started, though in fact Laplace had nothing to say about how the universe came into existence in the first place. He had not really eliminated the need for a creator, just for a god who interfered with things as they went along.

For many, though, this picture is not particularly comfortable, because they want a god with whom they have a relationship. In Laplace's mechanical universe there was no requirement for outside interference – though again there was nothing to stop a god taking a hand, just as we can reach out to a clock and change the position of its hands, overriding the clockwork.

As for randomness itself, we have seen how Einstein complained that God did not play dice. At one level this is a confusing statement as there is no evidence that Einstein had a religious faith. It seems that he used the term 'God' more as a compact way of saying 'the rules of how the universe works'. However, those who do believe in God and want their creator to be fair and just are sometimes unhappy with the unfair, uncaring nature of a universe that is so dependent on chaos and fundamental randomness at the quantum level.

Perhaps the best approach if you have a religious faith is to take a step back and say that it is entirely possible to separate the mechanism and the outcome. Just as all the everyday things we use, from computers to cars – and for that matter people – are made up of quantum particles that have weird, random behaviour but at the macro level carry on in a stable, non-random fashion, so we could envisage a creation that has elements of randomness and chaos but that seen at the macro level follows a creator's plan.

I am not saying that we can deduce there is a creator. If you like you can (and some have) make use of Bayesian statistics to try to determine whether or not God exists, but that is a futile exercise because we have no way of even approximating the probabilities we need to feed into the equation. Instead, what I am saying is that you pays your money and you takes your choice. The random foundations of the universe and the impact of chaos neither prevent nor encourage a belief in God.

When you look back over the way that our world manages to bring together such wild randomness and yet to appear stable and sometimes predictable, I think the best response is to have a sense of wonder. The more we discover about our dice world, the more fascinating it is. For me it's a major part of what makes science such fun. If physics was just about Newton's laws and the like it would do the job and give us good engineering, but it would be boring in the way that many people remember science being at school. But such people are wrong. And the reason they are wrong is that real science is so intriguing, mind boggling and fascinating, thanks to the dice world in which we live.

And what of free will? Perhaps the best we can do is to decide that in principle it doesn't exist, but the sensation of it being there is so impossibly real that we might as well think of it as being true. Free will, arguably, is the same as the steady, reliable existence of a chair. In principle it's a mass of random, weirdly behaving quantum particles that don't even stay in one place at a time. But in practice you can sit on it and it will stop you falling to the ground.

That's my impression of free will. But the final decision on whether or not it exists is up to you.

Or is it?

Index